Carotenoids - New Perspectives and Application

Edited by Rosa María Martínez-Espinosa

Published in London, United Kingdom

Carotenoids - New Perspectives and Application
http://dx.doi.org/10.5772/intechopen.95144
Edited by Rosa María Martínez-Espinosa

Contributors
Arslan Ahmad, Sakhawat Riaz, Muhammad Shahzaib Nadeem, Umber Mubeen, Khadija Maham, Daniel Glossman-Mitnik, Juan Frau, Norma Flores-Holguín, Susana M. Paixão, Ana S. Fernandes, Tiago P. Silva, José C. Roseiro, Luís Alves, Ivan M. Petyaev, Maria Margarida Cortês Vieira, Rose Daphnee Ngameni Tchonkouang, Maria Dulce Carlos Antunes, Lovina I. Iyandu Udoh, Eberechi R. Keyagha, Itorobong I. Nkanga, Josephine U. Agogbua, Rosa María Martínez-Espinosa

First published in London, United Kingdom, 2022 by IntechOpen
IntechOpen is the global imprint of INTECHOPEN LIMITED, registered in England and Wales, registration number: 11086078, 5 Princes Gate Court, London, SW7 2QJ, United Kingdom

British Library Cataloguing-in-Publication Data
A catalogue record for this book is available from the British Library

Additional hard and PDF copies can be obtained from orders@intechopen.com

Carotenoids - New Perspectives and Application
Edited by Rosa María Martínez-Espinosa
p. cm.

This title is part of the Physiology Book Series, Volume 16
Topic: Plant Physiology
Series Editor: Tomasz Brzozowski
Topic Editor: Jen-Tsung Chen

Print ISBN 978-1-80355-423-5
Online ISBN 978-1-80355-424-2
eBook (PDF) ISBN 978-1-80355-425-9
ISSN 2631-8261

IntechOpen Book Series
Physiology
Volume 16

Aims and Scope of the Series

Modern physiology requires a comprehensive understanding of the integration of tissues and organs throughout the mammalian body, including the cooperation between structure and function at the cellular and molecular levels governed by gene and protein expression. While a daunting task, learning is facilitated by identifying common and effective signaling pathways mediated by a variety of factors employed by nature to preserve and sustain homeostatic life. As a leading example, the cellular interaction between intracellular concentration of Ca+2 increases, and changes in plasma membrane potential is integral for coordinating blood flow, governing the exocytosis of neurotransmitters, and modulating gene expression and cell effector secretory functions. Furthermore, in this manner, understanding the systemic interaction between the cardiovascular and nervous systems has become more important than ever as human populations' life prolongation, aging and mechanisms of cellular oxidative signaling are utilised for sustaining life. Altogether, physiological research enables our identification of distinct and precise points of transition from health to the development of multimorbidity throughout the inevitable aging disorders (e.g., diabetes, hypertension, chronic kidney disease, heart failure, peptic ulcer, inflammatory bowel disease, age-related macular degeneration, cancer). With consideration of all organ systems (e.g., brain, heart, lung, gut, skeletal and smooth muscle, liver, pancreas, kidney, eye) and the interactions thereof, this Physiology Series will address the goals of resolving (1) Aging physiology and chronic disease progression (2) Examination of key cellular pathways as they relate to calcium, oxidative stress, and electrical signaling, and (3) How changes in plasma membrane produced by lipid peroxidation products can affect aging physiology, covering new research in the area of cell, human, plant and animal physiology.

Meet the Series Editor

Prof. Dr. Thomas Brzozowski works as a professor of Human Physiology and is currently a Chairman at the Department of Physiology and is V-Dean of the Medical Faculty at Jagiellonian University Medical College, Cracow, Poland. His primary area of interest is physiology and pathophysiology of the gastrointestinal (GI) tract, with a major focus on the mechanism of GI mucosal defense, protection, and ulcer healing. He was a postdoctoral NIH fellow at the University of California and the Gastroenterology VA Medical Center, Irvine, Long Beach, CA, USA, and at the Gastroenterology Clinics Erlangen-Nuremberg and Munster in Germany. He has published 290 original articles in some of the most prestigious scientific journals and seven book chapters on the pathophysiology of the GI tract, gastroprotection, ulcer healing, drug therapy of peptic ulcers, hormonal regulation of the gut, and inflammatory bowel disease.

Meet the Volume Editor

Rosa María Martínez-Espinosa is a Full Professor of Biochemistry and Molecular Biology at the University of Alicante, Spain, and has been the vice president of International Relations and Development Cooperation at this university since 2010. She created the research group in applied biochemistry in 2017 (https://web.ua.es/en/appbiochem/), and from 1999 to the present has made more than 200 contributions to Spanish and international conferences. Furthermore, she has around seventy-five scientific publications in indexed journals, eighty book chapters, and one patent to her credit. Her research work focuses on microbial metabolism (particularly on extremophile microorganisms), purification and characterization of enzymes with potential industrial and biotechnological applications, protocol optimization for genetically manipulating microorganisms, gene regulation characterization, carotenoid (pigment) production, and design and development of contaminated water and soil bioremediation processes by means of microorganisms. This research has received competitive public grants from the European Commission, the Spanish Ministry of Economy and Competitiveness, the Valencia Region Government, and the University of Alicante.

Contents

Preface

Carotenoids are colorful, natural, and versatile secondary metabolites that are widespread among living beings where they are involved in many biological roles reporting beneficial actions. During the last half-century, significant advances in carotenoids research have been made to improve knowledge about their composition, structure, and biological roles as well as to design more efficient processes for their production at a mid-large/large scale (involving new innovative approaches). Potential uses of carotenoids in different biotechnological fields like food science, biomedicine, and cosmetics are also being investigated.

Even so, there are certain limitations that make many of these carotenoid production processes inefficient and economically unprofitable. Examples of some of these limitations include the corrosion of bioreactors used to produce carotenoids at an industrial scale, for example, caused by certain conditions in highly ionic media; in other cases, the isolation of carotenoids from natural sources involves the use of organic reagents and solvents that are contaminants.

This book includes seven chapters addressing the role of carotenoids in certain diseases as well as in thermal adaptation in animals and plants. Chapters also explore new natural sources from which carotenoids showing high antioxidant capacities can be efficiently isolated and overproduced. Consequently, this book contributes to the dissemination of knowledge related to natural carotenoids characterization, their production at an industrial scale, and their potential uses in biomedicine among other fields. The examples collected herein are of interest not only to professionals and researchers in the field but also to students at the advanced undergraduate and graduate levels.

The editor gratefully thanks all the authors as well as the staff at IntechOpen for their valuable contributions. This book is designed to be a valuable source of new information on the many and varied facets of carotenoids.

Rosa María Martínez-Espinosa
Full Professor of Biochemistry and Molecular Biology,
Faculty of Science,
University of Alicante,
San Vicente del Raspeig, Alicante, Spain

Section 1

Introduction

Chapter 1

Introductory Chapter: Overview of New Perspectives and Applications of Carotenoids

Rosa María Martínez-Espinosa

1. Introduction

Carotenoids are naturally occurring terpenoid pigments made of isoprene residues, displaying a conjugated double polyene chain. They are hydrophobic compounds and are usually formed by a skeleton of 40 hydrocarbons, although most of the derivatives present in extremophilic microbes like those belonging to the Archaea domain contain 50 carbon atoms. This chain can be terminated by rings or functional groups with oxygen [1]. These compounds can absorb light mainly in the range of 300–600 nm (they show colors that vary from yellow to red) and are responsible for the characteristic pigmentation shown by some flowers, fungi, and several microorganisms. The ability to absorb light at a certain wavelength is related to the number of functional groups and double bonds present in their structure [2]. They are also involved in many biological roles in all living beings, reporting beneficial actions.

To date, over 1100 natural carotenoids have been described in animals, plants, macroalgae, some fungi, and a wide variety of microorganisms (including extremophiles). Due to the structural diversity of these molecules, there are still several biochemical and physiological functions to be associated with this class of compounds. Accordingly, these characteristics enable wide applicability, which drives the global carotenoid market.

Carotenoids are beneficial for human health; however, humans cannot synthesize carotenoids *de novo*, consequently, they are mainly obtained through the diet. In fact, carotenoids are consistently found in tissues or biological fluids, where they play a beneficial role in decreasing the risk of developing some diseases, such as cancer, eye disorders, and autoimmune or cardiovascular diseases.

During the last half-century, significant advances in the research of carotenoids have been made. As an example, much knowledge on their biosynthesis in plants and microbes has been generated, and there have been important breakthroughs in their production in both conventional and genetically modified organisms, not only in the laboratory but also on a large scale [3, 4].

The aim of this book is to highlight new perspectives and applications of carotenoids, including characterization and isolation of new compounds (including rare carotenoids), their production at a mid-large scale (involving new innovative approaches), and uses of carotenoids in different biotechnological fields, such as food science, biomedicine, and cosmetics.

2. Current research on carotenoids

Most of the recent studies on carotenoids describe the importance, chemical composition, and functioning of carotenoids. Although the research is mainly paying attention to C_{40} carotenoids, the interest in rare carotenoids, such as marine C_{50}, is increasing due to their significant antioxidant activity [3, 5].

There is also an important research field in which efforts are focused on looking for new natural sources from which natural carotenoids can be isolated. Considering their biological roles (not fully described yet), the potential applications of these molecules are still increasing. Some of the better described biological roles of carotenoids in plants, animals, and microbes are given below:

- Light absorbance mechanisms during photosynthesis.
- Protection of chlorophyll molecules and other cellular structures (not only in plants but also in animals, fungi, and microorganisms) from oxidative stress and reactive oxygen species (ROS) damage.
- Cellular signaling in plants: responses to environmental stresses, pollination, germination and reproduction, and development regulation.
- Strong antioxidant activity is primarily linked to their polyene molecular structure. Thus, carotenoids are reported as immune-enhancement and anticancer agents, which are also involved in the prevention of eye-, gastric and neurocognitive disorders, and in the regulation of obesity and anti-aging.
- Hormone precursors and precursors of vitamin A.

Among carotenoids, β-carotenes, astaxanthin, canthaxanthin, and lutein are highly marketed because they are part of drugs and cosmetics formulas or can be added to food due to their color or their antioxidant activities. So far, the large-scale production and subsequent commercialization of these compounds are carried out through chemical production, which has advantages and disadvantages compared to natural production. Thus, carotenoids chemically produced are characterized by excellent purity and consistency, with a relatively low production cost. However, the chemical synthesis of carotenoids involves the use of reagents, which are not environmentally friendly. Besides, the synthesis of pigments of greater structural complexity is expensive and highly time consuming. Considering all these disadvantages, the production of natural pigments using cell factories reveals a promising approach to obtaining carotenoids on a large scale [6]. In fact, carotenoids productions based on cell factories and environmentally friendly procedures is one of the fields of research attracting more attention worldwide during the last decade.

Recently, several works stated that carotenoids are also interested in sustainable energy and green electronics, which sheds light on more potential applications for these natural compounds [7].

3. Challenges related to carotenoid research and applications

Considerable progress in carotenoid research has been made to understand the carotenoid metabolism in all life forms, including humans. However, despite the extensive use of carotenoids in agri-foods sectors as well as in pharmaceutical industries, these pigments are an excellent example of poor understanding of food structure, the complexity of behavior during their digestion by animals, and inter-individual differences in response, which lead to misinterpretation of study results. Consequently, there are a few challenges associated with understanding and measuring carotenoid bioavailability, which should be the aim of research during the next few years. Some of the most relevant challenges are following, which are grouped into two categories—challenges related to the use of carotenoids to promote positive effects on human health and challenges related to the production of natural carotenoids at a large scale by environmentally friendly procedures:

Carotenoids and human health [8]:

i. To improve the knowledge about molecular mechanisms involved in the release of carotenoids from food structure and processing into an absorbable form (bioaccessibility). Recently, technologies of micro- and nanoencapsulation have addressed the needs of carotenoid entrapping to enhance their bioavailability, solubility, and chemical stability, thus ensuring the target delivery and the maintenance of their strong antioxidant among other biological activities [9].

ii. To characterize the transport and mobilization mechanisms of carotenoids from the gut lumen into the body (absorption). Absorption studies are best carried out by measuring chylomicron carotenoid excursion, with modeling of chylomicron turnover rate. In this way, inter-individual differences in lipoprotein metabolism could be considered before formulating conclusions on the rate and extent of absorption.

iii. Monitoring and interpreting plasma responses because of carotenoid digestion.

iv. To understand how inter-individual variation determines the final effect of carotenoid ingestion. Metabolomics and metabonomics reveal powerful tools to address this issue as well as those summarized in points ii) and iii) [10].

Production of natural carotenoids at a large scale by environmentally friendly procedures [11]:

i. To design sustainable, cheaper, and environmentally friendly protocols, which produce natural carotenoids at a large scale. This mostly refers to the need for the identification and cloning of genes responsible for carotenoid biosynthesis and transformation and related development of transgenic carotenoid-rich microorganisms or crops.

ii. Among standard and some advanced analytic tools for carotenoid isolation and characterization (e.g., high-performance liquid chromatography – HPLC, liquid chromatography–mass spectrometry – LC–MS, ultra-high-performance liquid chromatography – UHPLC, high-performance thin-layer chromatography – HPTLC, and others), nuclear magnetic resonance (NMR), Fourier-transform infrared spectroscopy (FTIR), or the primarily Raman spectroscopy coupled with chemometric modeling, opened a new era in carotenoid research and application. However, studies based on those techniques are still scarce.

Finally, one of the most challenging issues is the extraction of carotenoids from natural sources. So far, several extraction methods have been employed for the extraction of carotenoids—solvent extraction, Soxhlet extraction, centrifugation, and non-conventional methods of extraction, such as ultrasound-assisted, microwave-assisted, enzymatic, and the innovative technique supercritical carbon dioxide (SC–CO_2) extraction.

However, apart from SC–CO_2 extraction, which extracts pure compounds in high yield without the use of harmful organic solvents, the other mentioned approaches (which are the most efficient procedures to isolate pure carotenoids at high yield) are characterized by the use of large amounts of organic solvents. Consequently, green technology for carotenoid extraction is still the need of the present time to guarantee the production of marketed carotenoids for keeping safe the environmental quality for the next generations.

Acknowledgements

The author is thankful to Generalitat Valenciana, Spain (PROMETEO/2021/055) and the University of Alicante, Spain (VIGROB-309).

Conflict of interest

The author declares no conflict of interest.

Author details

Rosa María Martínez-Espinosa[1,2]

1 Faculty of Sciences, Biochemistry and Molecular Biology Division, Agrochemistry and Biochemistry Department, University of Alicante, Alicante, Spain

2 Multidisciplinary Institute for Environmental Studies "Ramón Margalef", University of Alicant, Alicante, Spain

*Address all correspondence to: rosa-martinez@ua.es

References

[1] Jehlicka J, Edwards HG, Oren A. Bacterioruberin and salinixanthin carotenoids of extremely halophilic Archaea and bacteria: A Raman spectroscopic study. Spectrochimica Acta. Part A, Molecular and Biomolecular Spectroscopy. 2013;**106**:99-103. DOI: 10.1016/j.saa.2012.12.081

[2] Johnson E, Schroeder W. Microbial carotenoids. In: Fiechter A, editor. Advances in Biochemical Engineering/Biotechnology. Springer: Berlin; 1996. pp. 119-178

[3] Torregrosa-Crespo J, Montero Z, Fuentes JL, Reig García-Galbis M, Garbayo I, Vílchez C, et al. Exploring the valuable carotenoids for the large-scale production by marine microorganisms. Marine Drugs. 2018;**16**(6):203. DOI: 10.3390/md16060203

[4] Srivastava A, Kalwani M, Chakdar H, Pabbi S, Shukla P. Biosynthesis and biotechnological interventions for commercial production of microalgal pigments: A review. Bioresource Technology. 2022;**352**:127071. DOI: 10.1016/j.biortech.2022.127071

[5] Saito T, Miyabe Y, Ide H, Yamamoto O. Hydroxyl radical scavenging ability of bacterioruberin. Radiation Physics and Chemistry. 1997;**50**:267-269

[6] Naziri D, Hamidi M, Hassanzadeh S, Tarhriz V, Maleki-Zanjani B, Nazemyieh H, et al. Analysis of carotenoid production by *Halorubrum*. sp. TBZ126: An extremely Halophilic Archeon from Urmia Lake. Advanced Pharmaceutical Bulletin. 2014;**4**:61-67. DOI: 10.5681/apb.2014.010

[7] Logoab AS, Mukherjee T. Application of carotenoids in sustainable energy and green electronics. Materials in Advance. 2022;**3**:1341-1358. DOI: 10.1039/D1MA01070K

[8] Rao AV, Rao LG. Carotenoids and human health. Pharmacological Research. 2007;**55**:207-216

[9] Faulks RM, Southon S. Challenges to understanding and measuring carotenoid bioavailability. Biochimica et Biophysica Acta. 2005;**1740**(2):95-100. DOI: 10.1016/j.bbadis.2004.11.012

[10] Arathia BP, Raghavendra-RaoSowmyaa P, Baskaranb KVV, Lakshminarayana R. Metabolomics of carotenoids: The challenges and prospects – A review. Trends in Food Science and Technology. 2015;**45**:105-117

[11] Singh A, Ahmadb S, Ahmad A. Green extraction methods and environmental applications of carotenoids-a review. RSC Advances. 2015;**5**:62358-62393

Section 2

Natural Sources of Carotenoids

Chapter 2

New Insights on Carotenoid Production by *Gordonia alkanivorans* Strain 1B

Tiago P. Silva, Susana M. Paixão, Ana S. Fernandes, José C. Roseiro and Luís Alves

Abstract

Gordonia alkanivorans strain 1B is a desulfurizing bacterium and a hyper-pigment producer. Most carotenoid optimization studies have been performed with light, but little is still known on how carbon/sulfur-source concentrations influence carotenoid production under darkness. In this work, a surface response methodology based on a two-factor Doehlert distribution (% glucose in a glucose/fructose 10 g/L mixture; sulfate concentration) was used to study carotenoid and biomass production without light. These responses were then compared to those previously obtained under light. Moreover, carbon consumption was also monitored, and different metabolic parameters were further calculated. The results indicate that both light and glucose promote slower growth rates, but stimulate carotenoid production and carbon conversion to carotenoids and biomass. Fructose induces higher growth rates, and greater biomass production at 72 h; however, its presence seems to inhibit carotenoid production. Moreover, although at a much lower yield than under light, results demonstrate that under darkness the highest carotenoid production can be achieved with 100% glucose (10 g/L), ≥27 mg/L sulfate, and high growth time (>216 h). These results give a novel insight into the metabolism of strain 1B, highlighting the importance of culture conditions optimization to increase the process efficiency for carotenoid and/or biomass production.

Keywords: Gordonia alkanivorans strain 1B, biomass, carotenoids, dark/light, glucose/fructose

1. Introduction

Carotenoids are bioactive molecules characterized by their intense coloration, which varies between red, yellow, orange, and pink, depending on their molecular structure. These molecules serve several biological purposes, aiding in photosynthesis in autotrophic organisms, protecting cells from excess light, have high antioxidant properties, and help in regulating membrane fluidity [1]. Due to these characteristics, carotenoids have garnered the interest of the health, cosmetic, pharmaceutical, and food industries, amongst others, giving them a high market value [2]. Indeed, the global carotenoids market was valued to grow from $1.5 billion in 2019 to $2.0 billion by 2026, recording an annual growth rate of 4.2% during the forecast period [3].

Carotenoids can be obtained from many sources, they can be chemically synthesized, extracted from plants and animals, or extracted from microorganisms. The latter option is becoming increasingly more appealing since microorganisms can be cultivated to higher densities, do not depend on seasonality, and can still be considered a natural source and a safe alternative to the synthetically derived pigments, which has become a deciding factor for consumers [4, 5].

Many microorganisms depend on light as a fundamental factor for carotenoid production. In the case of photosynthetic microorganisms, such as microalgae, known amongst the highest carotenoid producers, light is necessary for autotrophic growth and subsequent carotenoid production. Through photosynthesis, these microorganisms, consume CO_2 in the presence of light and use it to grow and produce byproducts. This dependence on light, however, leads to some constraints, since higher biomass concentrations will result in lower light penetration, which can hinder culture growth and reduce production yields [6]. As well, for heterotrophic growth, in microalgae, yeast, or bacteria, light can be fundamental as a powerful inducer, or necessary factor, for carotenoid production [7]. To overcome this problem, reactor designs had to be adjusted to increase surface area and light exposure, resulting in larger, more complex biorefineries, which demand higher extents of land use and/or greater initial costs [8].

When producing microbial biomass in an industrial setting, operation conditions can greatly impact production costs and process efficiency [9]. Many microbial-based industries struggle to optimize culture conditions to increase process efficiency [10]. In a biorefinery, carotenoid production could be viewed as a high-added-value byproduct and not as the focus of the process. In this perspective, the ability to produce carotenoids without a light source would make the process much easier, allowing the use of more conventional installations, and bioreactors, reducing the need for space, or complex infrastructures, leading to a reduction in installation costs [11].

The genus *Gordonia* is known for its carotenoid producers, such as strains of *Gordonia alkanivorans*, *Gordonia jacobea*, *Gordonia terrae*, *Gordonia ajoucoccus* and *Gordonia amicalis* [7, 12–15]. *Gordonia alkanivorans* strain 1B is a bacterium with high biotechnological interest. It has been extensively studied for its biodesulfurization (BDS) abilities, as a biocatalyst to substitute/complement the conventional fuel desulfurization methods. Using the 4S pathway, strain 1B can remove sulfur from complex organo-sulfur molecules, at ambient temperatures and pressures, without the need for additional treatments, potentially making the process more efficient and less pollutant [16–20]. However, one of the drawbacks that may hinder the BDS scale-up is the lack of economic viability, thus measures to reduce the overall process costs to make BDS with strain 1B economically competitive include the use of cost-effective feedstocks [21–24], culture medium minimization [10] and the exploitation of high-added value byproducts, such as biosurfactants and carotenoids [25, 26].

Carotenoid production is an attribute seldomly valorized in the literature related to biodesulfurizing microorganisms; however, it is commonly found in isolates from oil and oil-contaminated environments [27–31].

G. alkanivorans strain 1B was repeatedly described as a good carotenoid producer, presenting different concentrations and production profiles depending on its growth conditions [7, 25, 32]. Of the different carotenoids produced, three have been identified as canthaxanthin, astaxanthin, and lutein, by comparing with their respective standards through HPLC [7, 25]. Several carbon and sulfur sources have been tested as inducers, to increase carotenoid production with this strain, and initial studies have revealed that glucose and sulfate in abundance, in the presence of light, promote the highest accumulation of carotenoids [7, 32]. However, *Gordonia alkanivorans* strain 1B is one of the few described fructophilic bacteria [17], meaning it presents higher growth rates with fructose, producing biomass at faster rates, but also fewer carotenoids [32]. Furthermore, the presence of sulfate causes the inhibition of the biodesulfurization pathways. Concentrations as low as 30 mg/L almost completely inhibit desulfurization, even in the presence of organosulfur inducers [23].

Some work has already been performed to better understand how these factors (carbon-source/sulfur-source) correlate to generate the highest biomass and carotenoid productivity [32], however, it was mostly performed under the influence of light. Indeed, up to now, little is still known on how factors, such as carbon source and sulfur source concentrations, influence carotenoid production by *G. alkanivorans* strain 1B without the stimulus of light. Moreover, there is also a need to better understand the correlation between carotenoid accumulation, biomass production, and carbon consumption, with and without light. The correct balance between these responses is fundamental to better understand the metabolism of strain 1B and efficiently drive the process toward the production of either biomass (i.e., biocatalysts for desulfurization) or carotenoids, depending on the purpose of the biorefinery in consideration (bioproduct *versus* bioprocess).

This work initially focuses on the optimization of culture conditions toward carotenoid production by *G. alkanivorans* strain 1B without the stimulus of light. In this context, a surface response methodology (SRM) based on the Doehlert [33] distribution for two factors (% of glucose in a mixture of glucose + fructose (10 g/L total sugars); and sulfate concentration) was performed in the absence of light. Moreover, these SMR results (total biomass; total carotenoid production) were compared with the SRM results previously obtained by Fernandes et al. [32] for the carotenoids production in the presence of light (400 lux). In addition to biomass and carotenoids, specific carotenoid production (μg of carotenoids/g of dry cell weight), carbon consumption, and carotenoid and biomass production per carbon consumed were also evaluated as responses, both in absence/presence of light.

2. Materials and methods

2.1 Chemicals

Sodium sulfate anhydrous (>99%) was from Merck (New Jersey, USA). Dimethyl sulfoxide (DMSO) (99.9%), acetone (99.9%), ethyl acetate (99.8%), and methanol (99.9%) were obtained from Carlo Erba Reagents (Val de Reuil, France). The remaining reagents were of the highest grade commercially available. Stock solutions of glucose (glu) and fructose (fru) were prepared at 50% (w/v), filter sterilized, and stored for further use as a carbon source (C-source) in culture media. In the same way, a stock solution of 20 g/L Na_2SO_4 was also prepared and autoclaved (121°C, 1.03 bar, 15 min) to be further used as the sulfur source (S-source).

2.2 Microorganism and culture conditions

The microorganism used in this study was *G. alkanivorans* strain 1B, a bacterium isolated in our laboratory [16], and kept at a culture collection of microorganisms

(CCM at LNEG, Portugal, Lisbon). The basal salts medium used for cultivation, maintenance, and for all the growth/carotenoid production assays was described in Ref. [32]. The final pH was adjusted to 7.5 prior to sterilization by autoclave (121°C, 1.03 bar, 15 min). Afterward, the C-source (fructose and/or glucose) was added to the culture medium, in aseptic conditions, to an initial concentration of 10 g/L of total sugar(s). Similarly, the stock solution of S-source (Na_2SO_4) was also added to obtain the desired final concentrations of 9.04 mg/L, 22 mg/L, and 34.99 mg/L, depending on the assay.

The bacterial cultures were performed in 500 mL Erlenmeyer shake-flasks containing 150 mL culture medium, covered in tin foil to avoid light exposure, incubated in an orbital shaker (≈150 rpm) within an acclimatization chamber (Fitoclima 14000E Walk-In, Aralab, Portugal), at 30°C. All the assays were performed at least in duplicate. Sampling was carried out at 72 h and 216 h for immediate biomass determination (DCW = dry cell weight in g/L), while the remainder of each sample was centrifuged (8600 g at 4–5°C, 20 min in a refrigerated Sigma 2–16 K centrifuge) and the respective cells stored at −20°C until further pigment extraction and analysis. The supernatant was evaluated for sugar concentration through HPLC, using a Sugar-Pak 1 column (6.5 × 300 mm, 10 µm, Waters™, MA, USA) [32].

2.3 Experimental design methodology

A Doehlert distribution for two factors was used as the base for a surface response methodology (SRM) [33] to study carotenoid and biomass production by *G. alkanivorans* strain 1B in the absence of light (L0). The two factors studied were: X_1 – % of glucose in mixture glucose + fructose of 10 g/L of total sugars (0–100% glucose in the mix) and X_2 – sulfate concentration (7–37 mg/L of sulfate). Fourteen experiments (seven conditions in duplicate) were carried out. The results were evaluated in terms of responses (Y_i): biomass and total carotenoids production by strain 1B, at 72 h and 216 h. The model used to express the responses was a second-order polynomial model:

$$Y_i = \beta_0 + \beta_1 X_1 + \beta_2 X_2 + \beta_{12} X_{12} + \beta_{11} X_{12} + \beta_{22} X_{22} \quad (1)$$

where: Y_i – response from experiment i; β – parameters of the polynomial model; and X – experimental factor level [7, 23, 32].

In addition, specific carotenoid production (µg of carotenoids/g of DCW) was also evaluated as another response, both with and without a light source. The same polynomial model was applied to these results to generate the corresponding response surfaces. Model validation was performed through the Fischer test, for the effectiveness of the factors and the lack of fit, and R^2 (coefficient of multiple determination).

2.4 Carotenoid extraction and analysis

Carotenoid extraction and further characterization were performed following the procedure described in Ref. [7, 32]. The pigment results are presented as µg of total carotenoid produced (µg carotenoids per 150 mL) or as specific carotenoid production (µg of carotenoids/g of DCW).

3. Results and discussion

3.1 Experimental design (ED-L_0)

Table 1 shows the set of tests performed within the experimental design (ED), to study carotenoid and biomass production by strain 1B in the absence of light (L_0), and the responses obtained (biomass and total carotenoid production, both after 72 h and 216 h). **Figure 1** shows the response surfaces for the variation of biomass (**Figure 1A** and **B**) and total carotenoid production (**Figure 1C** and **D**), within the experimental domain, based on the responses observed.

In terms of biomass production, it is possible to see that at 72 h both factors influenced the response. The highest results were obtained with glucose ≤50% and sulfate at 22 mg/L. Further increases in glucose % resulted in a significant reduction of biomass regardless of sulfate concentration, with the lowest value being registered with 100% glucose, having a difference of more than six-fold when compared to the best results (0.72 vs. 4.66 g/L). This indicates that, at this time, when glucose % is above 50% it becomes the most influential factor, as corroborated by the more vertical lines of the left quadrants of the response surfaces (**Figure 1A**). Sulfate concentrations below and above 22 mg/L also resulted in lower biomass production,

Test (#)	Factors		Responses							
			Dark				Light*			
			Biomass (g/L)		Carotenoids (µg)		Biomass (g/L)		Carotenoids (µg)	
	Glucose (%)	Sulfate (mg/L)	72 h	216 h	72 h	216 h	72 h	216 h	72 h	216 h
1	50	22	3.87	3.53	74.6	110.2	2.42	4.2	230.8	400.5
2	50	22	4.17	3.76	56.2	125.5	2.30	4.65	215.4	435.8
3	100	22	0.81	4.06	40.6	336.8	0.32	4.17	103.7	912.7
4	100	22	0.63	3.90	39.7	285.3	0.40	3.27	82.6	675.6
5	0	22	4.06	3.25	31.8	92.6	2.86	4.55	290.2	462.0
6	0	22	5.25	3.05	40.3	69.0	2.84	4.63	303.1	493.8
7	75	34.99	3.39	3.79	52.8	185	2.42	5.00	286.8	592.5
8	75	34.99	3.11	3.59	61.3	171.4	2.80	4.57	315.4	514.4
9	25	9.01	1.99	2.19	5.1	34.0	1.83	2.39	104.0	136.3
10	25	9.01	2.87	2.13	11.0	31.6	1.75	2.30	195.2	257.0
11	75	9.01	2.55	2.45	19.4	58.7	1.81	2.43	151.6	204.2
12	75	9.01	2.93	2.30	31.6	45.7	1.79	2.27	155.8	197.7
13	25	34.99	3.52	3.23	20.3	84.4	2.48	4.71	225.0	427.0
14	25	34.99	3.99	2.99	21.1	87.1	1.99	4.71	235.3	427.0

**ED-$L_{400\ lux}$ results adapted/reprinted from Fernandes et al. [32]*

Table 1.
Doehlert distribution for two factors: % of glucose in mixture glucose + fructose (0–100%) and sulfate concentration (7–37 mg/L), and the responses evaluated (biomass and total carotenoids) in absence of light (ED-L_0) versus *with light (400 lux, ED-L_{400}). Seven conditions were tested in duplicates (14 tests), for statistical analysis.*

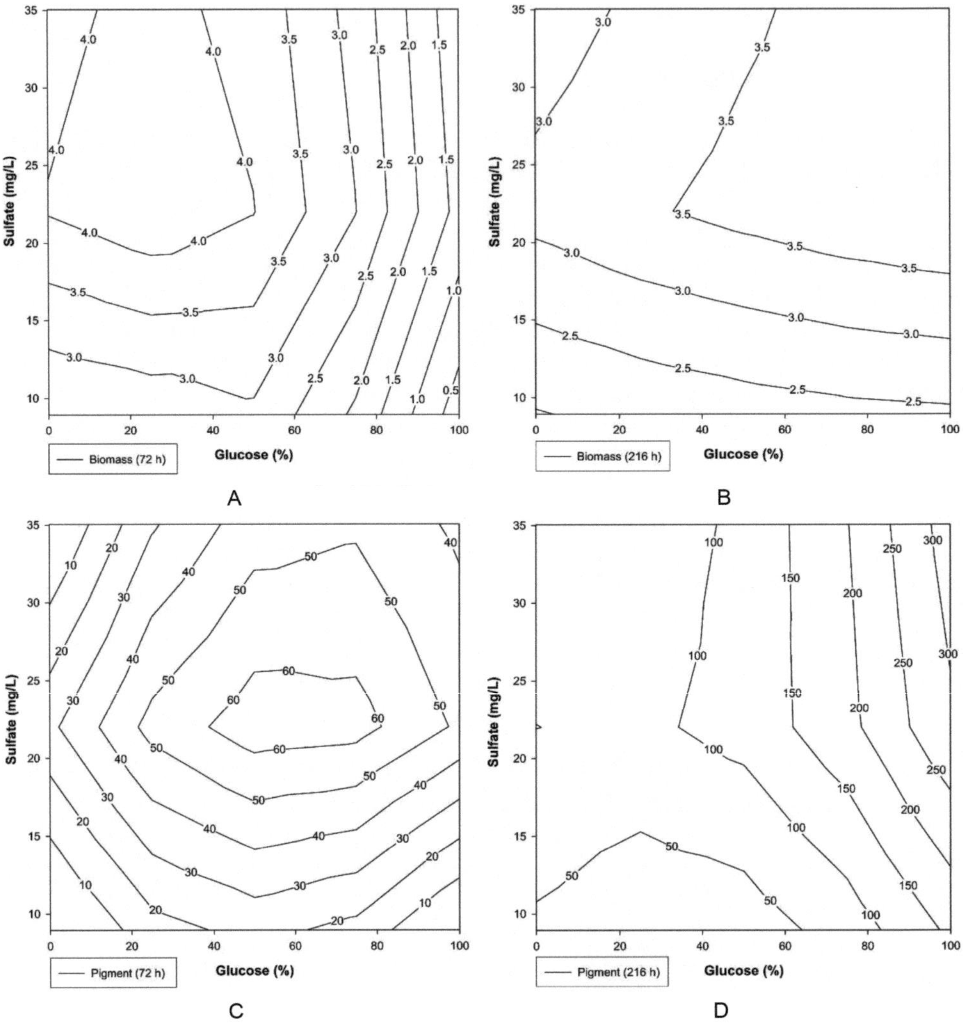

Figure 1.
Response surfaces for the biomass production (g/L) at 72 h (A) and 216 h (B); and for the total carotenoid production (µg) at 72 h (C) and 216 h (D), obtained in ED-L_0 for the factors % glucose in a mixture of fructose + glucose (0–100%) and sulfate concentration (7–37 mg/L).

even with glucose at 25%. This was especially evident with 9.01 mg/L of sulfate, as biomass never reached 3 g/L regardless of glucose concentration. In fact, sulfate was the most influential factor up to 20 mg/L, when glucose was below 50%, clearly demonstrated by the horizontal lines in the lower left quadrants of the response surfaces.

After 216 h, both factors continue to influence biomass production. However, contrary to what was observed at 72 h, an increase of glucose % was shown to have a positive effect on biomass, especially for values above 25%, regardless of sulfate concentration. The highest biomass concentration was observed with 100% glucose and 22 mg/L of sulfate (**Table 1**; tests 3–4: 3.98 g/L DCW), while the lowest results were registered with 9.01 mg/L of sulfate (**Table 1**; tests 9–10: 2.16 g/L DCW and tests 11–12: 2.38 g/L DCW). Increasing sulfate concentration resulted in a significant increase of biomass up to 22 mg/L, after which its influence is greatly reduced, regardless of glucose %, as seen in the lower half of **Figure 1B**.

The results obtained at both times (72 h and 216 h) are in accordance with the fructophilic nature of this strain, as described by Alves and Paixão [17]. Up to 72 h, when glucose and fructose are at a 50:50 ratio in a 10 g/L mix, there is no significant difference from growth with 0% glucose (100% fructose); therefore, by this time and within this range of glucose %, the growth is mostly finished, achieving the highest results. Further increases of glucose % result in lower biomass production, confirming that fructose has a stimulant effect on growth rates. When glucose represents 100% of the 10 g/L mix, and there is no fructose to induce biomass formation, the lag phase becomes longer, and the culture presents its lowest results. At 216 h, glucose seems to have a stimulant effect toward biomass production, however, this is also the result of its fructophilic nature. Between 72 h and 216 h, cultures with higher glucose % were at an earlier stage of their growth, and as such continued to increase biomass production. On the other hand, cultures with lower glucose %, which had already finished their development at 72 h, entered the stationary, or even cell death phase of the growth, resulting in stagnation, and/or reduction of biomass production, thus explaining this apparent contradiction.

Finally, it becomes evident that, under these conditions, a minimum sulfate concentration between 20 and 22 mg/L is needed to achieve significant biomass production. When sulfate was at 9.01 mg/L, biomass never reached a concentration of 3 g/L regardless of time or glucose %. On the other hand, an increase to 34.99 mg/L was mostly shown to have a small effect on biomass, thus reinforcing the notion that 22 mg/L of sulfate is sufficient for the consumption of 10 g/L of glucose/fructose.

In terms of total carotenoid production (μg of carotenoids per shake-flask), at 72 h, it is clear that both factors equally influence the response. In fact, at this time the optimum conditions were found at the center of the experimental domain. As seen in both **Table 1** and **Figure 1C**, this response is at its highest (65.4 μg) when glucose % is at 50%, and sulfate is of about 22 mg/L. Any significant deviation from these values results in a reduction of total carotenoid production.

At 216 h, total carotenoid production was positively influenced by both factors. Glucose % is shown to have the highest influence, increasing its impact for higher values of this parameter. This is especially true above 75%, since an increase to 100% glucose in the sugar mix resulted in an increase of 75% of total carotenoid production, from 178.2 to 311.1 μg, the highest observed in this ED. The lowest values were observed for conditions with sulfate at 9.01 mg/L (**Table 1**; tests 9-10: 32.8 μg and tests 11-12: 52.2 μg). As seen with biomass, increasing sulfate up to 22 mg/L resulted in an increase in response to more than double. Further increases in sulfate concentration have a smaller but positive impact. **Figure 1D** indicates that this production could be further increased by combining the maximum of both factors with glucose % at 100% and sulfate at 37.0 mg/L reaching a value close to 350 μg per culture flask.

At 72 h, it is possible to see equilibrium between biomass production and carotenoid induction, guaranteed by the presence of both sugars and enough sulfate to ensure complete carbon consumption. While glucose stimulates pigment production, it induces slower biomass formation, so if there is no fructose in the sugar mix, there will be fewer cells, resulting in less carotenoids.

Changing any of these conditions would result in a reduction of response, as demonstrated by the concentric lines of the response surface (**Figure 1C**). After 216 h of growth, the response changes and both factors presented a positive influence on the total carotenoid concentration, reaching a theoretical maximum at 100% glucose and

37 mg/L of sulfate. Overall, for lower glucose % in mix and/or lower sulfate concentration, sulfate concentration appeared to have greater influence, as seen in horizontal and diagonal lines in the lower quadrants. As for higher sulfate and glucose concentrations, the percentage of glucose/total sugars presented the highest influence, evidenced by the much more vertical lines in the upper quadrants.

3.1.1 Analysis of ED factors

The data obtained from the ED-L_0 was further used for regression analysis, and the polynomial model-derived parameters (β_0–β_{22}) are shown in **Table 2**. The β parameters of this polynomial model used to estimate the responses have the following meanings: β_0 represents the center of the experimental domain; β_1 and β_2 indicate the importance of the respective factors (factor 1: % glucose in a mixture glu + fru or glucose ratio, and factor 2: sulfate concentration, respectively) on the responses. The interaction parameter, β_{12}, indicates how the effect of one factor is dependent on the level of the other factor. β_{11} and β_{12} values determine how the response surface folds downward (negative values) or upward (positive values) quadratically, more or less rapidly in accordance with the magnitude of the absolute value [23].

	Environmental conditions	**Dark (L_0)**				**Light (L_{400})***			
	Responses	**Biomass**		**Carotenoids**		**Biomass**		**Carotenoids**	
	Time	**72 h**	**216 h**	**72 h**	**216 h**	**72 h**	**216 h**	**72 h**	**216 h**
Model parameters	β_0	4.02	3.65	65.4	117.87	2.36	4.43	223.14	418.21
	β_1	−1.34	0.41	10.33	95.39	−0.77	−0.1	−55.33	137.27
	β_2	0.53	0.65	12.76	51.66	0.36	1.08	65.8	150.83
	β_{12}	−0.47	0.21	10.91	42.17	0.21	0.65	38.59	105.36
	β_{11}	−1.33	−0.08	−27.3	78.06	−0.76	−0.27	−28.23	217.83
	β_{22}	−0.86	−1.06	−41.01	−66.86	−0.08	−1.43	−9.94	−191.0
Model validation (Fischer test)	Effectiveness of parameters	4.04	73.95	5.64	25.5	2.89	4.17	2.47	15.88
	Significance level (α), F (5,8)	0.04	0.001	0.02	0.001	0.09	0.04	0.12	0.001
	Lack of fit	23.68	0.02	17.25	17.17	93.51	3.29	35.82	0.78
	Significance level (α), F (1,7)	0.001	>0.100	0.004	0.004	0.001	>0.1	0.0005	>0.1
R^2	Coefficient of multiple determination	0.72	0.98	0.78	0.94	0.64	0.72	0.61	0.91

ED-$L_{400\ lux}$ results adapted/reprinted from Fernandes et al. [32]

Table 2.
Parameters of the polynomial model representing the responses studied (biomass production; total carotenoid production), with and without light (L_{400} vs L_0), at 72 h and 216 h. β_0, response at the center of the experimental domain; β_1 and β_2, parameters of the factors 1 (% glucose in a mix glu + fru) and 2 (sulfate concentration, mg/L), respectively; β_{12}, parameter of the interaction of the factors 1 and 2; β_{11} and β_{22}, self-interaction parameters of the factors 1 and 2, respectively.

At 72 h, β_1 and β_2 have opposite influences. β_1 presents the greatest value, indicating that glucose % has the highest influence on biomass production. However, being negative, β_1 also indicates that increasing this factor leads to a decrease in response, meaning that an increase of glucose % leads to a decrease in biomass. On the other hand, β_2 has a smaller, positive value indicating that an increase in sulfate concentration leads to an increase in biomass production. Analyzing pigment production at 72 h, β_1, β_2, and β_{12} presented positive values, indicating that the increase of each factor, individually or simultaneously, results in an increase of the response. Sulfate concentration was the factor with the greatest influence on pigment production.

At 216 h, there was a change in the response, as illustrated by the β parameters. Increasing each factor led to an increase in both biomass and pigment production. As shown in **Table 2**, for these conditions, sulfate concentration had a greater influence on biomass production (β_2 was 1.5-fold higher than β_1), however, pigment production was mostly influenced by glucose ratio (β_1 almost two-fold higher than β_2).

3.1.2 Comparing light and dark influence on carotenoid production

In a previous work [32], a similar experimental design was performed to evaluate the influence of light (400 lux). The results from that work, referred to as ED-L_{400}, are presented in **Table 1** and **Figure 2A–D**. From the comparison of ED-L_0 and ED-L_{400} results (**Figure 1** *versus* **Figure 2**), several differences become evident. In terms of biomass production, at 72 h optimum conditions do not differ substantially; however, the average results obtained under dark conditions were higher for every condition tested. This seems to indicate an inhibitory effect of the light source on growth rates, possibly resulting from the allocation of nutrients toward carotenoid production.

At 216 h, when sulfate concentration was 9.01 mg/L (**Table 1**: tests 9–12), there was no difference between light and dark cultures, and biomass did not surpass 3 g/L, reinforcing the observation that sulfate, at this concentration, was limiting. For the remaining conditions, there seemed to be a response in which cultures were grown with lower glucose %, and consequently with higher fructose concentrations, have lower biomass under dark conditions (**Table 1**: tests 5-6 and 13-14). As suggested above, this can be explained by the faster growth of these cultures under dark conditions, that by 216 h were already undergoing the cellular death phase. The inhibitory effect of the light source lowered the overall growth rates, to the point that, under light conditions, at 216 h these cultures were at an earlier stage of the growth, explaining the increased absorbance. For cultures with higher levels of glucose, since the growth rate is already slower, the inhibitory effects of light were not so evident with this two-point sampling (72 h and 216 h).

In terms of total carotenoid production, in both cases, the best results were obtained after 216 h with 100% glucose and 22 mg/L sulfate. However, the best average results obtained under light (~794 μg) were more than two-fold higher than those obtained in the dark conditions (~311 μg). Regardless of the presence of a light source, there was an increase of total carotenoid production from 72 h to 216 h for every condition tested. In both studies, at 72 h, this response was negatively influenced by higher glucose percentages in the mix. This effect, which is especially evident under light, seems to be mitigated, in both cases, by a higher concentration of sulfates. At 216 h this effect is reversed, and response is stimulated by glucose, since, as explained above, glucose induces slower growth rates, and longer division times would benefit such cultures.

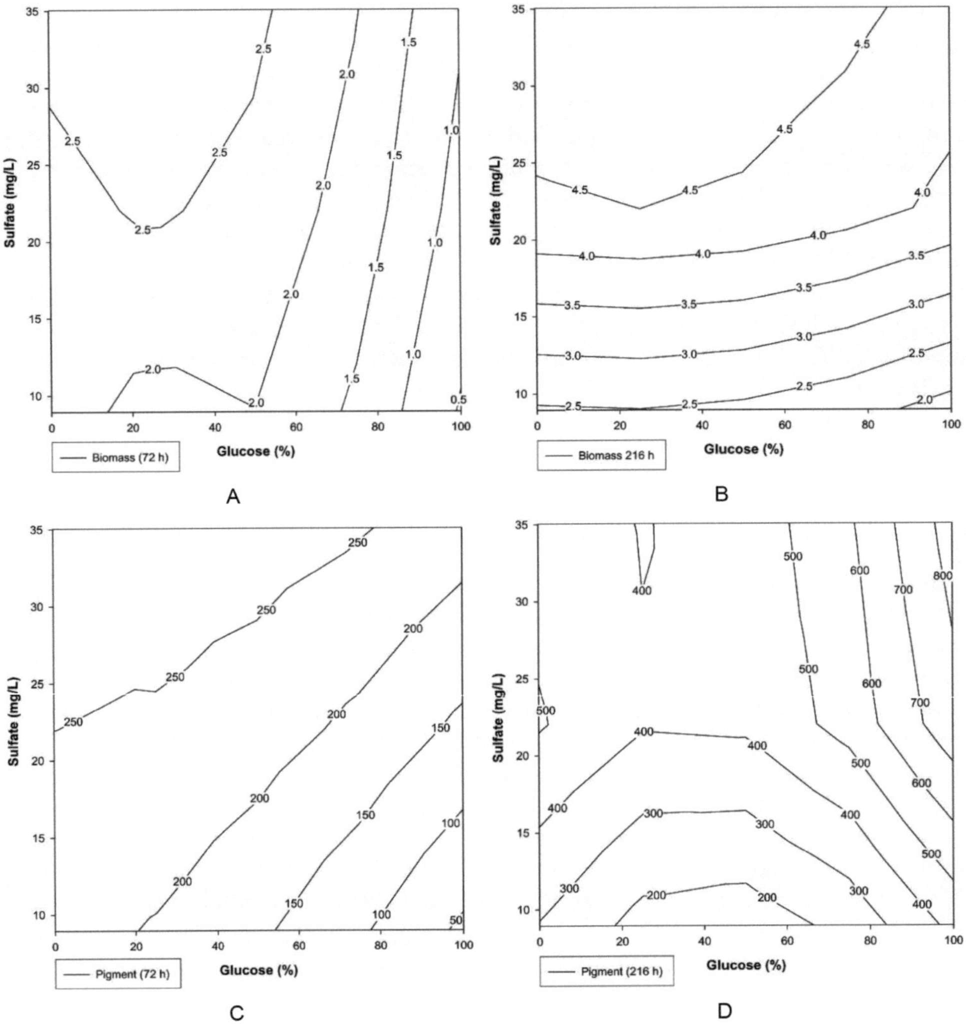

Figure 2.
Response surfaces for the biomass production (g/L) at 72 h (A) and 216 h (B); and for the total carotenoid production (μg) at 72 h (C) and 216 h (D), obtained in ED-L_{400} for the factors % glucose in a mixture of fructose + glucose (0–100%) and sulfate concentration (7–37 mg/L). Reprinted from Fernandes et al. [32].

Analyzing the beta parameters for both EDs (**Table 2**), the differences are again evident. At 72 h, in terms of biomass, each factor individually influences the response in a similar manner (β_1 and β_2). However, when both factors were increased simultaneously (β_{12}), the responses observed were opposite. Under dark conditions, this led to a decrease in biomass, while with light, it increased biomass production. This could indicate that, in the presence of light, the inhibitory effect that glucose has on biomass production can be partially reversed by increasing sulfate concentration, while under dark conditions, it can only be mitigated, maintaining the negative effect. The existence of this inhibitory effect was not previously observed, since, up to now, the works performed with sugar mixtures, did not take into account the conjugation of a lack of sulfur and light sources [17]. Nevertheless, Silva et al. [7] have already referred that strain 1B showed lower growth rates under the light. At 216 h the concentration of glucose has opposite effects depending on the presence of the light source. In the dark, it has a positive effect, while with light, it has a slightly negative influence. Furthermore, the relative influence of sulfate (β_2) was higher with light, while glucose (β_1) was higher without.

In terms of total carotenoid production, at 72 h, comparing β_1, β_2, and β_{12} to their respective β_0, it becomes clear that the studied factors presented a greater relative influence under light. Moreover, the increase of the glucose % (β_1) led to opposite responses. In the absence of light, an increase of glucose % resulted in an increase of total carotenoids, while in its presence, it greatly reduced total carotenoid production. At 216 h, total carotenoid production was positively influenced by all factors in both cases. However, all factors, especially glucose concentration, have greater relative importance under dark conditions. This indicates that in the absence of the stimulus of light, sulfate concentration, % of glucose in the sugar mix, and even time have greater importance for this response.

3.2 Specific carotenoid production

Total carotenoid production is an important parameter for the industrial process. Since it indicates the amount of carotenoids obtained in a certain volume of culture, it is deeply influenced by biomass production. The best conditions for total carotenoid production are obtained when there is a compromise between the highest carotenoid and highest biomass production. However, to better understand the mechanisms that influence carotenoid synthesis, it is fundamental to analyze specific carotenoid production (μg of carotenoids per g of dry cell weight - $\mu g_{(Carotenoids)}/g_{(DCW)}$ or $\mu g/g_{(DCW)}$). By evaluating the concentration of carotenoids per g of cells, it is possible to determine, which conditions induce greater cellular accumulation.

Using the results obtained in terms of biomass and total carotenoids, for each condition tested, the specific carotenoid concentrations were calculated, for both EDs, at dark and light conditions, at 72 h and 216 h and are presented in **Table 3**, with the corresponding response surfaces represented in **Figure 3A–D**.

Observing these results, it is possible to see that there are two very different behaviors in relation to time. Under the effects of light, the specific production of carotenoids is almost unchanged throughout the time period, from 72 h to 216 h (**Table 3**; **Figure 3A** and **B**). However, when the culture is grown in the darkness, time clearly has a significant influence, leading to a considerable increase of the carotenoids concentration present in the bacterial biomass, as well as to different influences of the studied factors, easily seen in the response surfaces in **Figure 3C** and **D**. In previous works, this feature had gone unnoticed since it is not observable when analyzing only the total carotenoid production.

From **Table 3**, it is also possible to determine that, for the same tests, specific carotenoid production was always greater in the presence of light, although this difference was more evident at 72 h. Tests 3 and 4, with glucose at 100% of the sugar mix and 22 mg/L of sulfates, always presented the highest response in each set of assays (L_{400}/L_0).

Further analyzing the results obtained in the presence of light, it is possible to see that, at both times, the two factors had a positive influence on specific carotenoid production. Maximum values were observed with 100% glucose and 22 mg/L of sulfate, while the lowest was registered with 9.01 mg/L of sulfates regardless of glucose percentage. As shown in **Figure 3A** and **B**, glucose % is the most influential factor, evidenced by the almost vertical lines, with sulfate having a smaller influence,

Test (#)			Response			
	Factors		Dark (L_0)		Light (L_{400})	
	Glucose (%)	Sulfate (mg/L)	72 h	216 h	72 h	216 h
			μg/g(DCW)	μg/g(DCW)	μg/g(DCW)	μg/g(DCW)
1	50	22	128.51	208.12	635.81	635.71
2	50	22	89.85	222.52	624.35	624.80
3	100	22	334.16	553.04	2160.42	1459.15
4	100	22	420.11	487.69	1376.67	1377.37
5	0	22	52.22	189.95	676.46	676.92
6	0	22	51.17	150.82	711.50	711.02
7	75	34.99	103.83	325.42	790.08	790.00
8	75	34.99	131.40	318.29	750.95	750.40
9	25	9.01	17.09	103.50	378.87	380.20
10	25	9.01	25.55	98.90	743.62	744.93
11	75	9.01	50.72	159.73	558.38	560.22
12	75	9.01	71.90	132.46	580.26	580.62
13	25	34.99	38.45	174.20	604.84	604.39
14	25	34.99	35.25	194.20	788.27	604.39

Table 3.
Doehlert distribution for two factors: % of glucose in mixture glucose + fructose (0–100%) and sulfate concentration (7–37 mg/L), and the response in terms of specific carotenoid production (μg/g(DCW)), with and without light (L_{400} vs L_0), at 72 h and 216 h. seven conditions were tested in duplicates (14 tests), for statistical analysis.

mostly for values below 22 mg/L. Optimum conditions for maximum specific carotenoid production are glucose at 100% of the sugar mix and sulfate concentration of at least 22 mg/L. These results also seem to reinforce the idea that the presence of fructose interferes with the cellular accumulation of carotenoids. Even at lower fructose concentrations (2.5 g/L in 10 g/L mix), specific carotenoid concentration only reaches 770 μg/g(DCW); however, when glucose represents 100% of 10 g/L mix, cellular accumulation of carotenoids has an increase of 140% and 94% at 72 h and 216 h, respectively, showing that a small concentration of fructose can have significant negative impacts on carotenoid production.

Under dark conditions, the responses obtained at both times were similar, but differences between tests were more pronounced at 72 h. As in previous results, the highest responses were obtained with glucose at 100% and 22 mg/L of sulfate (377 μg/g(DCW)). While the lowest results were observed with glucose at 25% at either sulfate concentrations (21.3 and 36.9 μg/g(DCW)), followed by glucose at 0% (51.7 μg/g(DCW)). This indicates a lower influence of the factors, when fructose is the dominant sugar, as demonstrated by the left quadrants of **Figure 3C**, which show a reduced variation of the response, regardless of glucose % and sulfate concentration. When glucose % is above 50%, both factors influence the response, with glucose showing the highest influence, as shown by the almost vertical lines on the right quadrants. Sulfate concentration is mostly important up to 22 mg/L, with higher values showing reduced or

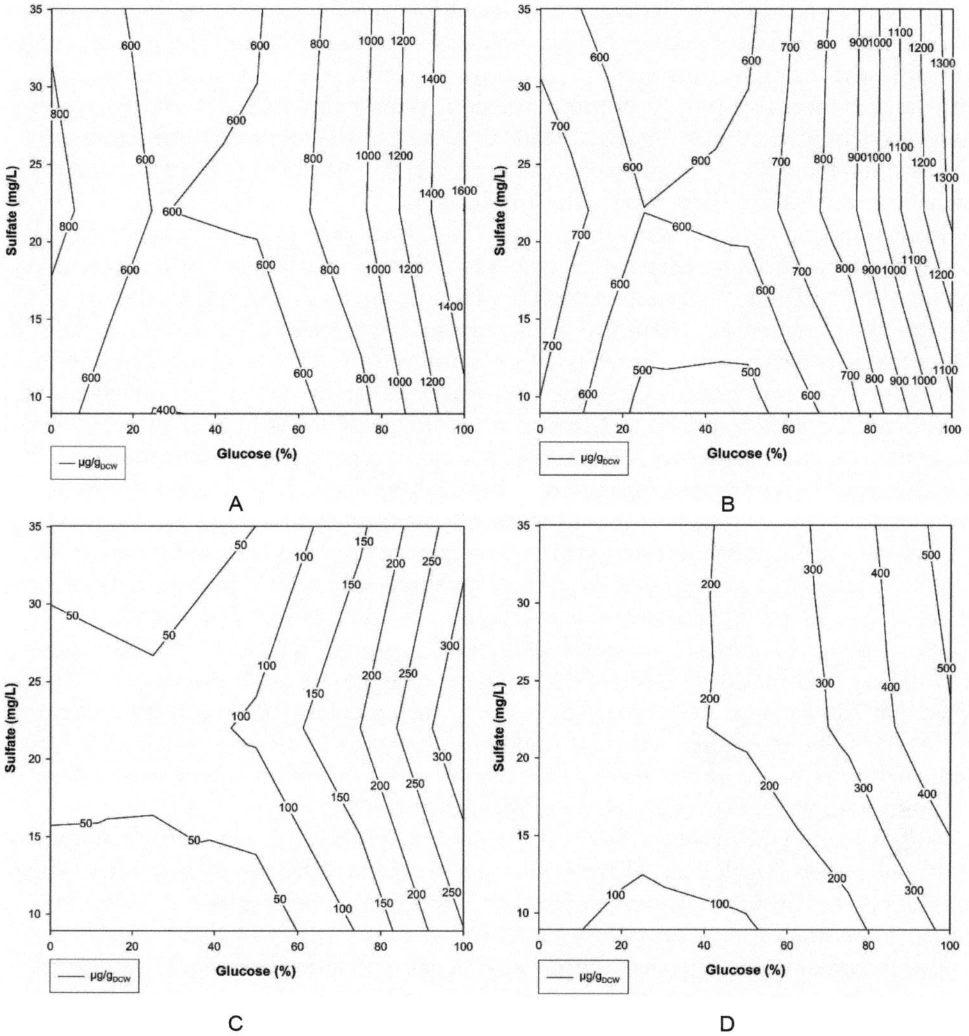

Figure 3.
Response surfaces for the specific carotenoid production (μg/g$_{(DCW)}$) with a light source (400 lux), at 72 h (A) and 216 h (B); and in the absence of light, at 72 h (C) and 216 h (D), obtained in ED-L$_0$ and ED-L$_{400}$ for the factors % glucose in a mixture of fructose + glucose (0–100%) and sulfate concentration (7–37 mg/L).

even negative impact on specific carotenoid production. So, **Figure 3C** shows that maximum cellular carotenoids would be obtained with 100% glucose and 22 mg/L, as it was tested (**Table 3**, tests 3 and 4).

At 216 h, carotenoid concentrations were higher for every condition tested. The highest value was recorded with 100% glucose and 22 mg/L of sulfate (520 μg/g $_{DCW}$). The lowest values were observed when sulfate was at 9.01 mg/L with glucose at 25% (101 μg/g$_{(DCW)}$) and 75% (146.10 μg/g$_{(DCW)}$). Both factors have a positive influence on the cellular accumulation of carotenoids; however, when sulfate is 22 mg/L or higher, glucose has the highest influence, shown by the vertical lines in the upper right corner of **Figure 3D**. Results indicate that the highest specific carotenoid production could be achieved with glucose 100% and sulfate >35 mg/L.

These results also indicate that a concentration of 9.01 mg/L of sulfate hinders specific carotenoid production, not only biomass production, and 22 mg/L of sulfate are sufficient for high carotenoid production, at earlier stages of the growth (72 h), with further increases having minimal impact. At later times of growth, increasing sulfate concentration to 34.99 mg/L leads to increases in cellular accumulation of carotenoids, most likely because cellular needs for growth are met and it can be diverted toward secondary metabolite production.

Most importantly, fructose seems to have an inhibitory effect on carotenogenesis. Under these conditions, since the tests were performed with a mix of fructose and glucose, with different percentages while maintaining 10 g/L of total sugars, when glucose % was increased, there was a proportional decrease in fructose. So, when glucose was increased from 25 to 50% (and sulfate from 9.01 to 22 mg/L) there was a 25% reduction in fructose, which caused a very significant increase in the accumulation of carotenoids (five-fold at 72 h and two-fold at 216 h). A further increase of both glucose and sulfate values to 75% and 34.99 mg/L, respectively, resulted in a much lower or nonexistent increase in the response, especially at 72 h. However, by completely removing fructose from the mix, from 50–0% fructose (**Table 3**; tests 1-2 and 5-6), even without increasing sulfate concentration, there was an increase in response ($\mu g_{(Carotenoids)}/g_{(DCW)}$) of more than three-fold at 72 h and more than two-fold at 216 h. This seems to indicate an inhibitory effect, which is slightly lower at 216 h, since most fructose has been completely consumed within 72 h to 76 h, possibly attenuating its inhibitory effects. Furthermore, cells grown with glucose at 25% were those that had the highest increase over time, raising cellular concentration of carotenoids 4.75- and five-fold; while for higher glucose %, the increase was under three-fold, further reinforcing the theory that carotenoid production is increased after fructose disappears, especially when glucose is present.

Sugars having inhibitory effects on carotenoid synthesis is a known phenomenon, mostly observed for glucose. Some researchers proposed the use of alternative sugars, or alcohols, as C-sources, to achieve higher carotenoid concentrations. In fact, a similar phenomenon was described for *Xanthophyllomyces dendrorhous*, a yeast that produces astaxanthin. This microorganism starts to produce carotenoids at the stationary phase of the growth if grown with glucose but starts the production at the beginning of the growth if cultivated with succinate [34]. Since *G. alkanivorans* strain 1B is a fructophilic bacterium [17], it could be the case that similar mechanisms are being applied to its preferred C-source, explaining the reduced carotenoid production in the presence of this sugar.

Table 4 presents the beta parameters for the polynomial model used for the specific carotenoid production, showing the influence of each factor studied (% glucose in a mix glu + fru; sulfate concentration). It is possible to note that both factors have a positive influence on the response. It is also clear that glucose percentage is the most influential factor on the conditions tested (**Figure 3**), being several times more influential than sulfate concentration in all cases.

Considering the responses at the center of the domain (β_0), it becomes clear that both factors have a greater relative influence under dark conditions, despite having significantly lower responses. Glucose concentration is especially influential at 72 h, as β_1 is greater than β_0. Moreover, under light, the relative importance of the individual factors reduces with time, while under dark conditions, the relative influence of sulfate increases, and that of the glucose percentage is reduced.

Comparing the overall results for specific carotenoid production (**Table 3**; **Figure 3**) with those obtained for total carotenoid production (**Table 1**, **Figure 2**), it

	Response	$\mu g_{(Carotenoids)}/g_{(DCW)}$			
	Environmental conditions	Dark (L_0)		Light (L_{400})	
	Time	72 h	216 h	72 h	216 h
Model parameters	β_0	109.19	215.35	630.19	630.37
	β_1	128.61	147.08	371.86	270.38
	β_2	20.74	74.69	97.14	69.74
	β_{12}	23.54	53.55	38.04	91.19
	β_{11}	105.23	130.03	601.09	424.76
	β_{22}	−101.63	−79.38	−174.79	−146.61
Model validation (Fischer test)	Effectiveness of parameters	14.01	28.95	4.15	9.32
	Significance level (α), F(5,8)	0.001	0.001	0.05	0.01
	Lack of fit	19.22	18.11	5.86	9.9
	Significance level (α), F(1,7)	0.01	0.01	0.05	0.05
R^2	Coefficient of multiple determination	0.9	0.95	0.72	0.85

Table 4.
Parameters of the polynomial model representing specific carotenoid production ($\mu g/g_{(DCW)}$), with and without light, at 72 h and 216 h. β_0, response at the center of the experimental domain; β_1 and β_2, parameters of the factors 1 (% glucose in a mix glu + fru) and 2 (sulfate concentration, mg/L), respectively; β_{12}, parameter of the interaction of the factors 1 and 2; β_{11} and β_{22}, self-interaction parameters of the factors 1 and 2, respectively.

becomes visible that, under light, cellular carotenoid production is not influenced by time above 72 h, meaning that total carotenoid production increased due to an increase in biomass concentration and not in the cellular carotenoid concentration. In the same way, the sulfate concentration had a greater influence on total carotenoid production, due to its importance for biomass production. Values above 22 mg/L had little effect on cellular carotenoid production, despite increasing total carotenoid production. Under dark conditions, by comparing the response surfaces obtained for specific carotenoid production (**Figure 3C** and **D**) with those obtained for total carotenoids (**Figure 1C** and **D**), it is possible to observe that there are significant differences at 72 h since at this time the biomass produced is very different depending on the conditions tested. So, the conditions for the highest total carotenoid production will lead to more biomass, but not the cells at the highest pigmentation. At 216 h, the figures are much similar, since the differences between biomass were much smaller.

3.3 Influence of the carbon source

3.3.1 Carbon consumption

Sugar consumption was greatly affected by the influence of light, under some conditions, sugars were fully consumed, while in others sugar consumption was residual. Due to this range of results, carbon consumption could not be properly represented by the models previously applied. However, given their importance to understanding the metabolic response, total sugar consumption results were displayed in **Figure 4A–D**, to better illustrate how they were influenced by the factors studied in EDs (L_0 and L_{400}).

Under dark conditions, at 72 h (**Figure 4C**), it is possible to see that the tests with more glucose and less sulfate were those in which sugar consumption was the lowest. Except for the conditions when glucose was at 75%, fructose was always completely consumed, which seems to indicate a slower consumption of glucose, expected from this fructophilic strain. At 216 h (**Figure 4D**), sugars were completely consumed, apart from the tests with lower sulfate concentration, which maintained the same concentration values presented at 72 h, and one of the replicates with 100% glucose. This reinforces the concept that sulfate levels were limiting the growth of the culture, and that a value higher than 9.01 mg/L is needed to consume 10 g/L of sugars.

Under light, at 72 h, none of the tests resulted in the complete consumption of sugars. The highest result (8.81 g/L) was observed when glucose was at 0% and fructose was the sole C-source. As shown in **Figure 4A**, increasing glucose % resulted in a decrease in sugar consumption that was especially evident above 50% glucose with 22–34.99 mg/L of sulfate. The lowest sugar consumption was observed when glucose was 100% of the 10 g/L sugar mix, where the culture was still in its lag phase, and sugar consumption was residual (0.0042 g/L). At 216 h (**Figure 4B**), sugars were fully consumed when glucose percentage was up to 50% and sulfate concentration was ≥22 mg/L. At 75% glucose with 34.99 mg/L sulfate, less than 2.2 g/L were left as residual, and at 100% glucose, less than 5 of the 10 g/L of initial glucose were consumed.

Comparing both light and dark results, there was a clear inhibitory effect of the light source, in terms of consumption of both sugars, this had only been described for glucose, mostly because assays with fructose ended before 72 h. Moreover, these results also point out that, with sufficient sulfur source, carbon source consumption is

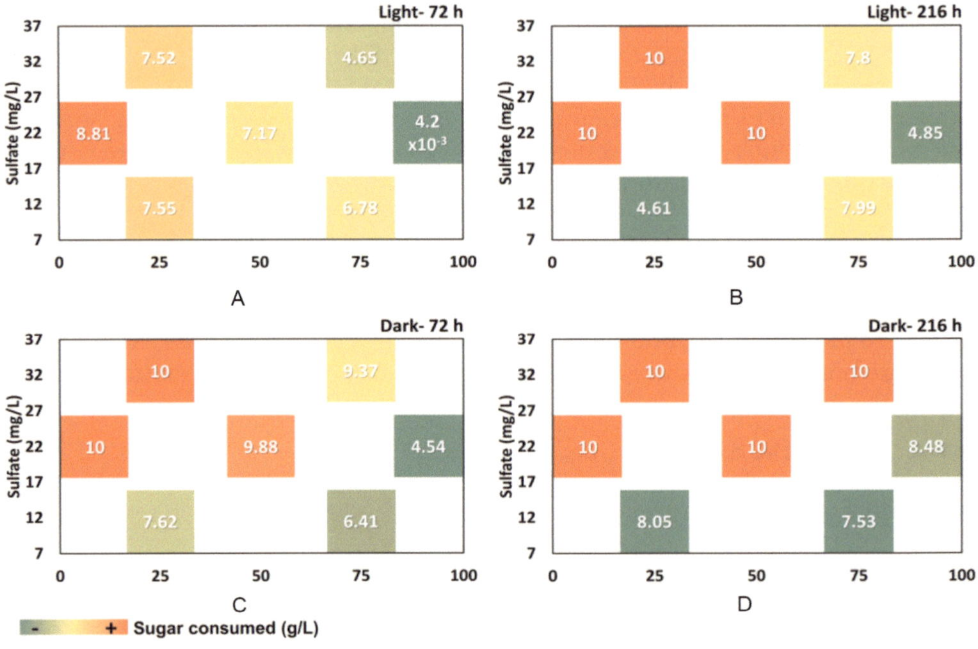

Figure 4.
Doehlert distribution for two factors: % of glucose in a mix glu + fru (0–100%) and sulfate concentration (7–37 mg/L), and the responses in terms of total sugar consumed (g/L), with light at 72 h (A) and 216 h (B) and in dark at 72 h (C) and 216 h (D), respectively. Seven conditions were tested in duplicates (14 tests), for statistical analysis.

reduced, there is an accentuated fructophilic behavior and a significant effect of the glucose concentration under light conditions. The presence of fructose, greatly accelerated sugar consumption, as seen when comparing the results of 75% glucose and 100% glucose on every assay.

3.3.2 Carbon conversion to biomass

To better understand and prepare a future biorefinery it is important to evaluate how carbon is diverted toward biomass production. As such, based on the results for biomass production and carbon consumption, **Figure 5A–D** was created, illustrating biomass produced (g/L DCW) per total sugar consumed (g/L), that is, (g/g), for the different conditions tested.

At 72 h, with light (**Figure 5A**), the highest result was achieved with 75% glucose and 34.99 mg/L of sulfate (0.52 g/g), any reduction in glucose % or sulfate concentration resulted in a decrease in response. The remaining conditions had values between 0.33 and 0.25 g/g. Results with glucose at 100% of the mix were disregarded, since as previously stated, under these conditions the culture was still in its lag phase, there was residual carbon consumption and residual biomass production. At 216 h (**Figure 5B**), carbon conversion to biomass increased on every condition tested, nonetheless, the general trends remained similar. The highest result was registered with 100% glucose (0.76 g/g), followed by 75% glucose with 34.99 mg/L of sulfate (0.65 g/g), the lowest was also observed with 75% glucose but with 9.01 mg/L (0.3 g/g). The remaining results varied between 0.44 g/g and 0.51 g/g.

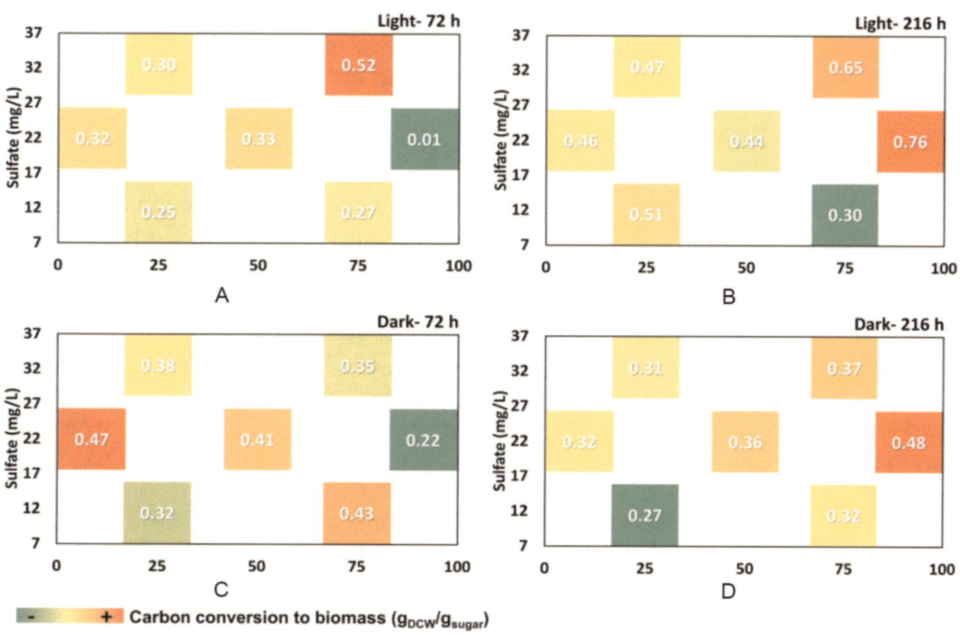

Figure 5.
Doehlert distribution for two factors: % of glucose in a mix gluc + fru (0–100%) and sulfate concentration (7–37 mg/L), and the responses in terms of biomass produced/sugar consumed (g/g), with light at 72 h (A) and 216 h (B) and in dark at 72 h (C) and 216 h (D), respectively. Seven conditions were tested in duplicates (14 tests), for statistical analysis.

This shows that higher concentrations of glucose and sulfate result in a higher carbon conversion to biomass. However, since fructose is needed to induce faster biomass production, 100% glucose leads to lower initial results, this can be surpassed by maintaining a small concentration of fructose, no more than 25% of total sugars. Furthermore, it also seems to indicate that carbon conversion to biomass increases with time. This could result from the accumulation of reserve substances at later times, in response to the stimulus of light, coupled with a slower metabolic activity which results in less carbon being converted to CO_2.

Under dark conditions, at 72 h (**Figure 5C**), significant differences were mostly observed at both extremes of glucose %, with 22 mg/L of sulfate. The highest result occurred with 0% glucose (0.47 g/g) and the lowest with 100% glucose (0.22 g/g). The remaining conditions resulted in values between 0.32 g/g and 0.43 g/g without a defined tendency. At 72 h sulfate appears to be less influential to how cells allocate carbon, most influence is centered on the presence/absence of fructose and glucose. Indicating that the nature of carbon source leads to different metabolic responses, which could be related to the fructophilic behavior of *G. alkanivorans* strain 1B. After 216 h (**Figure 5D**), the lowest results were observed with glucose <50% or sulfate <22 mg/L (0.27–0.32 g/g). The highest result was achieved with 100% glucose and 22 mg/L of sulfates (0.48 g/g). Overall, carbon conversion to biomass appears to be stimulated by higher glucose %. Sulfate concentration also stimulates this response, up to 22 mg/L, greater increases seem to have no influence. This could indicate that fructose stimulates the production of extracellular compounds, such as biosurfactants [26], and glucose the accumulation of reserve substances, such as lipids [35], thus diverting carbon to different pathways.

Comparing the results obtained, at 72 h, carbon conversion to biomass was mostly higher under dark conditions, however, at 216 h this is reversed. Moreover, higher glucose and sulfate concentrations at later times, also present higher carbon conversion rates to biomass, with the highest results being obtained at 216 h, with light, 100% glucose, and 22 mg/L of sulfate.

Without light, when there was lower glucose %, time led to a reduction in carbon conversion to biomass, even if there was very few or no carbon left at 72 h. With greater glucose percentages, there was an increase of carbon conversion to biomass from 72 h to 216 h, even when sugar was already consumed at 72 h. Comparing each individual value obtained at 216 h and 72 h (**Figure 5C** and **D**), it becomes clear that, over time, carbon conversion to biomass increases with the increase in glucose % (0 < 25 < 50 < 75 < 100%). This increase with time could be the result of the accumulation of reserve substances, such as lipids, sugars, or PHA's, induced by light and/or glucose, resulting in slower growth rates, and higher carbon conversion yields. Conversely, without the stimulus of light, especially at greater fructose concentrations, growth rates are higher, but there is a loss of biomass over time, which could indicate that cells produce fewer reserve substances, or that these are converted, or released into the medium in the form of biosurfactants or other exopolysaccharides, resulting in loss of dry weight. Alternatively, this could be the result of a higher metabolic activity induced by fructose, as highlighted by Alves and Paixão [17], which could result in cells with lower abundance of reserve substances, leading to greater cellular lysis over time.

Other fructophilic bacteria and yeast have been shown to convert fructose to mannitol, which is accumulated and used as an osmolyte, a carbon reserve, or an antioxidant, substituting carotenoids in the latter function. However, the same phenomenon is not observed when glucose is the C-source [36, 37]. A similar mechanism could be occurring here, in which fructose is converted into mannitol or another intermediary and then consumed for the cell metabolism, while glucose is consumed at a slower speed, and further converted to carotenoids and other reserve substances, which would be consumed after the 216 h evaluated in this work (**Figure 4**).

3.3.3 Specific carotenoid production per gram of sugar consumed

A final response was still obtained by combining the responses in terms of specific carotenoid production ($\mu g_{(Carotenoids)}/g_{(DCW)}$) per total sugar consumed (g/L), that is, specific carotenoid production per sugar consumed ($\mu g_{(Carotenoids)}/g_{(DCW)}/g_{(Sugars)}/L$). This combined response indicates how efficiently sugars are converted into carotenoids in each cell. The results obtained are illustrated in **Figure 6A–D**.

After 72 h under light, the lowest results were registered when glucose was at 0% or sulfate was 9.01 mg/L (72.31–84.17 $\mu g_{(Carotenoids)}/g_{(DCW)}/g_{(Sugars)}/L$). Increases in glucose % or sulfate concentration always resulted in an increased response. The highest result calculated was obtained for 75% glucose and 34.99 mg/L of sulfate (165.71 $\mu g_{(Carotenoids)}/g_{(DCW)}/g_{(Sugars)}/L$). It should be noted that cells grown on 100% glucose and 22 mg/L of sulfate revealed much greater specific carotenoid production, and since carbon consumption at this point was still residual, the calculated result was several thousand times greater than any other, losing most of its comparative meaning (**Figure 6A**). However, this is not devoid of biological meaning, since it indicates that under these conditions, before biomass production begins, or sugar consumption starts, carotenoid production is already occurring, and as seen above at the same specific concentration, as later times.

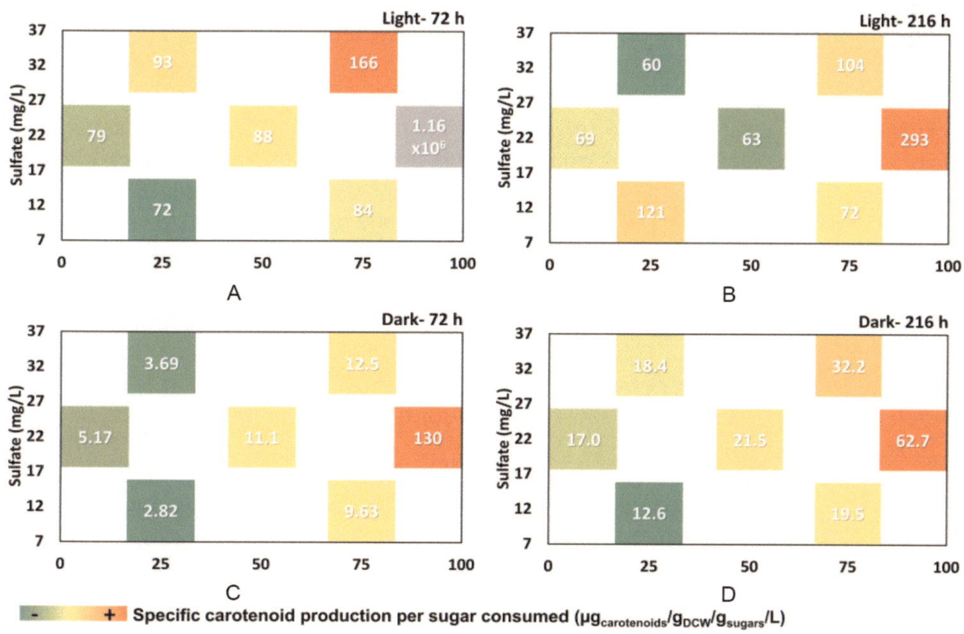

Figure 6.
Doehlert distribution for two factors: % of glucose in a mix glu + fru (0–100%) and sulfate concentration (7–37 mg/L), and the response in terms of specific carotenoid production per sugar consumed ($\mu g_{(Carotenoids)}/g_{(DCW)}/g_{(Sugars)}/L$), with light at 72 h (A) and 216 h (B) and in dark at 72 h (C) and 216 h (D), respectively. Seven conditions were tested in duplicates (14 tests), for statistical analysis.

After 216 h, as demonstrated in **Figure 6B**, the highest values were achieved when glucose was at 100% of the 10 g/L sugar mix and sulfate at 22 mg/L (293 $\mu g_{(Carotenoids)}/g_{(DCW)}/g_{(Sugars)}/L$). The lowest values were obtained when glucose % was between 25 and 50% with sulfate ≥22 mg/L (60.4–63 $\mu g_{(Carotenoids)}/g_{(DCW)}/g_{(Sugars)}/L$).

These results show that, at both times, glucose is the factor with most influence. At 72 h, the factors studied had a greater influence at higher concentrations of sulfate and glucose %, and both glucose and sulfate have a positive effect. At 216 h, the response presented two behaviors. For lower concentrations of sulfate, lower glucose % benefits the response, while for higher concentrations of sulfates, greater glucose % benefits the response. This could be due to the fact that lower concentrations of sulfate result in less biomass, leading to an excess carbon source, which could help induce the synthesis of carotenoids, as a response to the light. But, with greater sulfate concentration glucose acts as the inducer. Moreover, both at 72 h and 216 h, 100% glucose proved to be the best condition, even at lower sulfate concentration.

In the absence of light, at 72 h (**Figure 6C**), the highest results were obtained with glucose at 100% and 22 mg/L of sulfate, while the lowest were observed with glucose at 25% and sulfate at 9.01 mg/L (130 and 2.82 $\mu g_{(Carotenoids)}/g_{(DCW)}/g_{(Sugars)}/L$, respectively). At 216 h (**Figure 6D**), the maximum and minimum values were registered at the same conditions (62.7 and 12.6 $\mu g_{(Carotenoids)}/g_{(DCW)}/g_{(Sugars)}/L$, respectively). This clearly demonstrates that both factors have a positive effect on the response, also showing that under dark conditions, especially at 72 h, glucose is the most influential factor. At 216 h response values were higher, apart from the tests with 100% glucose; however, the relative importance of glucose % was lower, as shown by the closer response values, and sulfate gained influence, when glucose % was <100%. Furthermore, as previously stated, fructose seems to have an inhibitory effect on the carotenoid production, and on the efficiency of converting carbon into carotenoids, as evidenced by the difference in response values between 75% and 100% glucose, especially significant at 72 h (**Figure 6C**).

Comparing the results obtained for specific carotenoid production per sugar consumed ($\mu g_{(Carotenoids)}/g_{(DCW)}/g_{(Sugars)}/L$) at both times, with and without light (**Figure 6**), it becomes clear that time, for most conditions, had opposite influences on the response. With light, carbon conversion to carotenoids mostly decreased over time, while without light, it seems to increase, except when glucose is used at 100% of the sugar mix. This fact could be a conjugation of several different factors. At earlier stages of the growth, the cells were fewer in number and had greater exposure to light, due to a lower cell density. As the cultures progressed, the number of cells increased and there was an increase in shade effect. Since under dark conditions the stimulus of light is not present, the values are lower, but increase over time. Other factors, such as oxygen and nutrient concentration vary in a similar fashion in both assays, and as such, should play a smaller role in this difference. The result observed for 100% glucose at 72 h under light is an extreme example, where the culture was only starting to grow at that point, however, the cells were heavily stimulated to produce carotenoids. The cells present at this time were mostly from the initial inoculum, and they were induced to produce carotenoids, by the presence of light, and high concentrations of glucose (100% of the 10 g/L mix) and sulfate (22 mg/L), before starting to grow and consume sugar in a significant manner.

The presence of light enhances the conversion of carbon into carotenoids in each cell, as demonstrated by the smaller response under dark conditions. Moreover, carotenoid production seems to be induced by sugar but is not dependent on it. As stated above, with light, at 72 h and 100% glucose (10 g/L), the cells were extremely rich in carotenoids, but sugar consumption was minimal. This indicates that strain 1B could be producing the carotenoids based on cellular reserve substances present in the cells of the inoculum. This is reinforced by the fact that, in the dark, with 0% glucose, and 25% glucose, there is a complete sugar consumption by 72 h (**Figure 4C** and **D**), and the specific carotenoid production increases at 216 h (**Figure 6C** and **D**). This evidence confirms what had already been reported in the work by Silva et al. [7], in which strain 1B cells grown in the dark with fructose and DBT were placed in the light after the end of the growth and have developed coloration, without additional extracellular sugar.

4. Conclusion

This work reinforces the importance of the sulfate concentration, the nature and ratio of the sugars used as C-source, the presence of light, and the growth time for the production of biomass and carotenoids by *G. alkanivorans* strain 1B, leading to a better understanding of how these factors interact with each other to influence different metabolic responses.

In terms of sulfate, it shows that >20 mg/L are needed to consume 10 g/L C-source (glucose/fructose), while guarantying efficient biomass and carotenoid production. However, this concentration is enough to cause significant biodesulfurization inhibition, making it difficult to conjugate high carotenoid production with high desulfurization ability, without adapting the biocatalyst production method.

In terms of carbon, as expected, the presence of fructose leads to faster growth rates and greater biomass production. However, at later times, after growth has ended, it also led to a loss of biomass, especially in the absence of light, probably due to the production of extracellular compounds, such as biosurfactants. Moreover, fructose seems to inhibit carotenoid production to some extent, since even 25% of fructose can result in a great loss in carotenoid production. Glucose, on the other hand, hinders growth rates and stimulates carotenoid production and conversion of carbon to carotenoids and biomass, possibly indicating the accumulation of reserve substances, thus justifying the longer growth rates. The presence of light also seems to reduce growth rates, and stimulate carotenoid production, while making the fructophilic behavior more evident. In addition, the growth time period is especially important to generate specific carotenoids without light, and biomass in the presence of either light or greater glucose concentrations. All these responses seem to indicate that the fructophilic behavior of this strain is not simply a matter of sugar transport, since there are different metabolic behaviors in the presence/absence of both sugars.

The overall results indicate that higher glucose concentrations combined with more light, or a better-adapted system, with a higher surface to volume ratio, could drastically increase carotenoid production. In addition, this study confirms that is possible to produce carotenoids under dark conditions, and that production can be greatly stimulated by culture medium conditions. Moreover, it also reinforces that even without sugar consumption it is possible to induce carotenoid production in cells of strain 1B, opening the possibility of developing two-phase systems of biomass production based on fructose without light, and further carotenogenesis induction with light exposure, under optimal conditions.

Ultimately, these results may help in the development of a future biorefinery, either pointing the way to generate carotenoids under dark conditions, as an added value byproduct, or further increment carotenoid production in the light, as the main bioproduct.

Acknowledgements

This work was financed by national funds through FCT (Fundação para a Ciência e a Tecnologia) in the scope of the project GreenFuel (PTDC/EAM-AMB/30975/2017). Tiago P. Silva also acknowledges FCT for his PhD financial support (SFRH/BD/104977/2014).

Author details

Tiago P. Silva, Susana M. Paixão*, Ana S. Fernandes, José C. Roseiro and Luís Alves*
LNEG – National Laboratory of Energy and Geology, Unit of Bioenergy and Biorefineries, Estrada do Paço do Lumiar, Lisbon, Portugal

*Address all correspondence to: susana.alves@lneg.pt and luis.alves@lneg.pt

References

[1] Ram S, Mitra M, Shah F, Tirkey SR, Mishra S. Bacteria as an alternate biofactory for carotenoid production: A review of its applications, opportunities and challenges. Journal of Functional Foods. 2020;**67**:103867. DOI: 10.1016/j.jff.2020.103867

[2] Amengual J. Bioactive properties of carotenoids in human health. Nutrients. 2019;**11**:1-6. DOI: 10.3390/nu11102388

[3] McWilliams A. The Global Market for Carotenoids. Wellesley, Massachusetts, United States: BCC Research; 2018. Report ID: FOD025F

[4] Gong M, Bassi A. Carotenoids from microalgae: A review of recent developments. Biotechnology Advances. 2016;**34**:1396-1412. DOI: 10.1016/j.biotechadv.2016.10.005

[5] Barreiro C, Barredo J-L. Carotenoids Production: A Healthy and Profitable Industry. Berlin/Heidelberg, Germany: Springer; 2018. pp. 45-55. DOI: 10.1007/978-1-4939-8742-9_2

[6] Khan MI, Shin JH, Kim JD. The promising future of microalgae: Current status, challenges, and optimization of a sustainable and renewable industry for biofuels, feed, and other products. Microbial Cell Factories. 2018;**17**:1-21. DOI: 10.1186/s12934-018-0879-x

[7] Silva TP, Paixão SM, Alves L. Ability of *Gordonia alkanivorans* strain 1B for high added value carotenoids production. RSC Advances. 2016;**6**:58055-58063. DOI: 10.1039/c6ra08126f

[8] Gifuni I, Pollio A, Safi C, Marzocchella A, Olivieri G. Current bottlenecks and challenges of the microalgal biorefinery. Trends in Biotechnology. 2019;**37**:242-252. DOI: 10.1016/j.tibtech.2018.09.006

[9] Alves L, Paixão SM, Pacheco R, Ferreira AF, Silva CM. Biodesulphurization of fossil fuels: Energy, emissions and cost analysis. RSC Advances. 2015;**5**:34047-34057. DOI: 10.1039/C4RA14216K

[10] Pacheco M, Paixão SM, Silva TP, Alves L. On the road to cost-effective fossil fuel desulfurization by: *Gordonia alkanivorans* strain 1B. RSC Advances. 2019;**9**(44):25405-25413. DOI: 10.1039/c9ra03601f

[11] Hu J, Nagarajan D, Zhang Q, Chang JS, Lee DJ. Heterotrophic cultivation of microalgae for pigment production: A review. Biotechnology Advances. 2018;**36**:54-67. DOI: 10.1016/j.biotechadv.2017.09.009

[12] Kim JH, Kim SH, Yoon JH, Lee PC. Carotenoid production from n-alkanes with a broad range of chain lengths by the novel species *Gordonia ajoucoccus* A2T. Applied Microbiology and Biotechnology. 2014;**98**:3759-3768. DOI: 10.1007/s00253-014-5516-y

[13] Lee K-W, How CW, Chen L, Chen PT, Lan JC-W, Ng H-S. Integrated extractive disruption of *Gordonia terrae* cells with direct recovery of carotenoids using alcohol/salt aqueous biphasic system. Separation and Purification Technology. 2019;**223**:107-112. DOI: 10.1016/j.seppur.2019.04.031

[14] Veiga-Crespo P, Blasco L, do Santos FR, Poza M, Villa TG. Influence of culture conditions of *Gordonia jacobaea* MV-26 on canthaxanthin production. International Microbiology. 2005;**8**:55-58

[15] Sowani H, Deshpande A, Gupta V, Kulkarni M, Zinjarde S. Biodegradation of squalene and n-hexadecane by *Gordonia amicalis* HS-11 with concomitant formation of biosurfactant and carotenoids. International Biodeterioration and Biodegradation. 2019;**142**:172-181. DOI: 10.1016/j.ibiod.2019.05.005

[16] Alves L, Salgueiro R, Rodrigues C, Mesquita E, Matos J, Gírio FM. Desulfurization of dibenzothiophene, benzothiophene, and other thiophene analogs by a newly isolated bacterium, *Gordonia alkanivorans* strain 1B. Applied Biochemistry and Biotechnology. 2005; **120**(3):199-208. DOI: 10.1385/ABAB:120:3:199

[17] Alves L, Paixão SM. Fructophilic behaviour of *Gordonia alkanivorans* strain 1B during dibenzothiophene desulfurization process. New Biotechnology. 2014;**31**:73-79. DOI: 10.1016/J.NBT.2013.08.007

[18] Paixão SM, Arez BF, Roseiro JC, Alves L. Simultaneously saccharification and fermentation approach as a tool for enhanced fossil fuels biodesulfurization. Journal of Environmental Management. 2016;**182**:397-405. DOI: 10.1016/j.jenvman.2016.07.099

[19] Paixão SM, Silva TP, Arez BF, Alves L. Advances in the reduction of the costs inherent to fossil fuels' biodesulfurization towards its potential industrial application. Biotechnology. 2019;**22**:390-425. DOI: 10.4018/978-1-5225-8903-7.ch081

[20] Alves L, Melo M, Mendonça D, Simões F, Matos J, Tenreiro R, et al. Sequencing, cloning and expression of the dsz genes required for dibenzothiophene sulfone desulfurization from *Gordonia alkanivorans* strain 1B. Enzyme and Microbial Technology. 2007;**40**: 1598-1603. DOI: 10.1016/J.ENZMICTEC.2006.11.008

[21] Alves L, Paixão SM. Enhancement of dibenzothiophene desulfurization by *Gordonia alkanivorans* strain 1B using sugar beet molasses as alternative carbon source. Applied Biochemistry and Biotechnology. 2014;**172**(6):3297-3305. DOI: 10.1007/S12010-014-0763-Z

[22] Silva TP, Paixão SM, Roseiro JC, Alves L. Jerusalem artichoke as low-cost fructose-rich feedstock for fossil fuels desulphurization by a fructophilic bacterium. Journal of Applied Microbiology. 2015;**118**:609-618. DOI: 10.1111/JAM.12721

[23] Silva TP, Paixão SM, Teixeira AV, Roseiro JC, Alves L. Optimization of low sulfur carob pulp liquor as carbon source for fossil fuels biodesulfurization. Journal of Chemical Technology & Biotechnology. 2013;**88**:919-923. DOI: 10.1002/JCTB.3921

[24] Alves L, Marques S, Matos J, Tenreiro R, Gírio FM. Dibenzothiophene desulfurization by *Gordonia alkanivorans* strain 1B using recycled paper sludge hydrolyzate. Chemosphere. 2008;**70**: 967-973. DOI: 10.1016/j.chemosphere.2007.08.016

[25] Silva TP, Alves L, Paixão SM. Effect of dibenzothiophene and its alkylated derivatives on coupled desulfurization and carotenoid production by *Gordonia alkanivorans* strain 1B. Journal of Environmental Management. 2020;**270**: 110825. DOI: 10.1016/J.JENVMAN.2020.110825

[26] Tavares J, Alves L, Silva TP, Paixão SM. Design and validation of an expeditious analytical method to quantify the emulsifying activity during biosurfactants/bioemulsifiers

production. Colloids and Surfaces B: Biointerfaces. 2021;**208**:112111. DOI: 10.1016/j.colsurfb.2021.112111

[27] Lee DW, Lee H, Lee AH, Kwon BO, Khim JS, Yim UH, et al. Microbial community composition and PAHs removal potential of indigenous bacteria in oil contaminated sediment of Taean coast, Korea. Environmental Pollution. 2018;**234**:503-512. DOI: 10.1016/J.ENVPOL.2017.11.097

[28] Salehi Bakhtiyari A, Etemadifar Z, Borhani MS. Use of response surface methodology to enhance carotenoid pigment production from *Cellulosimicrobium* strain AZ. SN Applied Sciences. 2020;**2**:1-9. DOI: 10.1007/S42452-020-03549-6/TABLES/4

[29] Subhash Y, Lee SS. Roseomonas deserti sp. nov., isolated from crude oil contaminated desert sand. International Journal of Systematic and Evolutionary Microbiology. 2018;**68**:675-680. DOI: 10.1099/IJSEM.0.002565/CITE/REFWORKS

[30] Venil CK, Malathi M, Devi PR. Characterization of Dietzia maris AURCCBT01 from oil-contaminated soil for biodegradation of crude oil. Biotechnology. 2021;**11**:1-13. DOI: 10.1007/S13205-021-02807-7/TABLES/3

[31] Matsui T, Maruhashi K. Isolation of Carotenoid-deficient Mutant from Alkylated Dibenzothiophene Desulfurizing Nocardioform Bacteria, *Gordonia* sp. TM414. Current Microbiology. 2004;**48**(2):130-134. DOI: 10.1007/S00284-003-4141-2

[32] Fernandes AS, Paixão SM, Silva TP, Roseiro JC, Alves L. Influence of culture conditions towards optimal carotenoid production by *Gordonia alkanivorans* strain 1B. Bioprocess and Biosystems Engineering. 2018;**41**:143-155. DOI: 10.1007/s00449-017-1853-4

[33] Doehlert DH. Uniform Shell designs. Applied Statistics. 1970;**19**:231. DOI: 10.2307/2346327

[34] Wozniak A, Lozano C, Barahona S, Niklitschek M, Marcoleta A, Alcaíno J, et al. Differential carotenoid production and gene expression in *Xanthophyllomyces dendrorhous* grown in a nonfermentable carbon source. FEMS Yeast Research. 2011;**11**:252-262. DOI: 10.1111/j.1567-1364.2010.00711.x

[35] Gouda MK, Omar SH, Aouad LM. Single cell oil production by *Gordonia* sp. DG using agro-industrial wastes. World Journal of Microbiology and Biotechnology. 2008;**24**:1703-1711. DOI: 10.1007/s11274-008-9664-z

[36] Meena M, Prasad V, Zehra A, Gupta VK, Upadhyay RS. Mannitol metabolism during pathogenic fungal–host interactions under stressed conditions. Frontiers in Microbiology. 2015;**6**. Article 1019, (12 pages). DOI: 10.3389/fmicb.2015.01019

[37] Filannino P, di Cagno R, Tlais AZA, Cantatore V, Gobbetti M. Fructose-rich niches traced the evolution of lactic acid bacteria toward fructophilic species. Critical Reviews in Microbiology. 2019; **45**:65-81. DOI: 10.1080/1040841X.2018.1543649

Chapter 3

Potential of Carotenoids from Fresh Tomatoes and Their Availability in Processed Tomato-Based Products

Rose Daphnee Ngameni Tchonkouang,
Maria Dulce Carlos Antunes and Maria Margarida Cortês Vieira

Abstract

The high consumption of tomatoes worldwide has made them an essential source of health-promoting carotenoids that prevent a variety of chronic degenerative diseases, such as diabetes, high blood pressure, and cardiovascular disease. Tomatoes are available year-round, consumed fresh, and used as a raw material for the production of many processed products, such as juices, pastes, and purees. A plethora of carotenoids has been characterized in tomatoes. Most of the relevant carotenoids in the human bloodstream are supplied by fresh and processed tomatoes. Lycopene is the predominant carotenoid in tomato and tomato-based food products. Other carotenoids such as α-, β-, γ- and ξ-carotene, phytoene, phytofluene, neurosporene, and lutein are present in tomatoes and related products. There is a growing body of evidence that these bioactive compounds possess beneficial properties, namely anti-carcinogenic, cardioprotective, and hepatoprotective effects among other health benefits, due to their antioxidant, anti-mutagenic, anti-proliferative, anti-inflammatory, and anti-atherogenic properties. This chapter analyzes the carotenoid composition of tomatoes and their based products as major contributors to the chronic disease-preventive properties.

Keywords: tomato, carotenoids, health, antioxidant, chronic diseases

1. Introduction

Tomato (*Solanum lycopersicum* L.) belongs to the Solanaceae family and is one of the most important and versatile crops in the world. Wild tomato is originated in western South America along the coast and the high Andes from central Ecuador, through Peru, to northern Chile, and in the Galapagos Islands [1–5]. The period and center of domestication of tomatoes are still unclear, but it is believed that the region of domestication was Central America [1–3]. Nowadays, with several selected cultivars adapted to different edaphoclimatic conditions and final food products, tomato is cultivated worldwide and marketed either as fresh or as processing tomato [4]. The global tomato production reached 186,821,216 tons in 2020. According to FAOSTAT, the world's top three tomato producers in 2020 were China, India, and Turkey [6].

Tomato consumption is dominant in China, India, North Africa, the Middle East, the US, and Brazil [5]. They are consumed either fresh in salads or cooked in sauces, soup, and various dishes. In addition, they can be processed into paste, purées, juices, and ketchup. Canned and dried tomatoes are economically important processed products, they contain abundant minerals, essential amino acids, sugars, and dietary fibers [4] as well as many health-promoting compounds including vitamins, carotenoids, and phenolic compounds [5]. Although these bioactive constituents could be obtained through intake of dietary supplements, scientific evidence suggests that direct intake from their natural matrix is much more effective regarding health benefits [7].

Carotenoids are well-known bioactive compounds involved in preventing the development of diseases such as diabetes, gastrointestinal and cardiovascular diseases (CVDs), for example, by reducing the amount of oxidized low-density lipoproteins (LDLs). They are also known to reduce the risk of developing degenerative diseases such as blindness, xerophthalmia, and degeneration of muscles. In addition, carotenoids possess anticancer properties in health conditions, such as stomach, lung, and prostate cancers [3], being this disease-preventing action attributed to their antioxidant components. Lycopene and β-carotene are carotenoids with particularly strong antioxidant activities, based on their abilities to quench singlet oxygen and trap peroxyl radicals [8]. In this chapter, the potential of tomato carotenoids in chronic disease prevention is discussed. The role and types of carotenoids are presented, after which the composition and distribution of carotenoids in tomato and tomato-based products are documented. The factors influencing the bioavailability of tomato carotenoids are explained. Finally, the action of carotenoids in the risk reduction of non-communicable diseases is detailed.

2. Carotenoids: definition, classification, and food sources

Carotenoids represent a large family of non-water-soluble pigments that range from yellow to red and are predominantly found in fruits and vegetables [9, 10]. Generally, carotenoids are a class of isoprenoid molecules that are commonly referred to as pigments due to their characteristic yellow to red color. This physical property is due to a polyene chain containing 3–13 conjugated double bonds that act as a chromophore. All photosynthetic organisms (such as plants and algae) and some non-photosynthetic bacteria and fungi synthesize carotenoids that are tetraterpenes (terpenes consisting of eight isoprene units, $C_{40}H_{64}$) derived from phytoene, a 40-carbon isoprenoid [11, 12]. Some carotenoids called higher carotenoids are made up of a 45- or 50-carbon skeleton, while those having carbon skeletons with fewer than 40 carbons are called apocarotenoids [13]. Carotenoids can be synthesized de novo by flora and microbes, and do not occur naturally in mammals with minor exceptions [14]. Therefore, carotenoids found in animal tissues are either directly obtained from their diets or partially modified during metabolic reactions [15, 16]. Carotenoids are essential compounds in all photosynthetic species, such as algae, cyanobacteria, and plants, and are involved in basic physiological processes, such as photoprotection and photosynthesis. They serve numerous important functions, such as light-harvesting, photoprotection during photosynthesis, and photo-oxidative damage prevention, and also accumulate in non-photosynthetic organs of plants, such as fruits, pericarps, seeds, roots, and flowers. They provide color to flowers and fruits which is useful in pollination and seed dispersal through pollination vector attraction. They also serve as precursors for the biosynthesis of the phytohormone, abscisic acid (ABA) in non-photosynthetic organs [12, 13, 16].

2.1 Classification of carotenoids

Carotenoids can broadly be classified into two subgroups according to their chemical structure—(1) carotenes (hydrocarbon carotenoids), which are made up of carbon and hydrogen. Examples of carotenes include α-carotene, β-carotene, β, ψ-carotene (γ-carotene), and lycopene; (2) oxycarotenoids or xanthophylls (oxygenated carotenoids), which are derivatives of the hydrocarbons (carotenes) and are constituted by carbon, hydrogen, and oxygen atom in the form of hydroxy, epoxy, or oxy groups. Examples of xanthophylls include β-cryptoxanthin, lutein, zeaxanthin, astaxanthin, fucoxanthin, and peridinin [13, 14, 16–18]. Moreover, carotenoids are divided into primary or secondary. Primary carotenoids are compounds required by plants in photosynthesis (β-carotene, violaxanthin, and neoxanthin), whereas secondary carotenoids are localized in non-photosynthetic organs of plants, such as fruits and flowers (α-carotene, β-cryptoxanthin, zeaxanthin, antheraxanthin, capsanthin, and capsorubin) [14].

2.2 Carotenoids in foods

More than 700 naturally occurring carotenoids have been identified, and new carotenoids are continuously identified [17]. The nutritionally important carotenoids in human foods include the carotenes; β-carotene, α-carotene, and lycopene and the

Carotenoid	Food sources	Health properties
β-Carotene	Carrot, sweet potato, mango [19], pumpkin, kale, apricots, pepper, tomato paste [19], cassava [21]	Protection against oxidative stress due to the inactivation of reactive oxygen species (ROS) [22] Anticancer properties [22] Risk reduction of CVDs [23] Protects against macular degeneration and reduces aging [24]
α-Carotene	Spinach, cantaloupe [25] Tomato [26, 27]	Antioxidant and anticarcinogenic agent [23, 25]
Lycopene	Tomato and processed products, red carrot, red bell pepper, watermelon, papaya [19]	Decreased risk of prostate cancer [19] Strong antioxidant effect due to the inactivation of ROS and the quenching of free radicals [22] Anticancer properties [23] Risk reduction of CVDs [23] Reduce the risk of macular degeneration [24]
β-Cryptoxanthin	Oranges, papaya, peaches, tangerines, maize (yellow/orange) [28] mangoes [25], tomato [26, 27]	Antioxidants and anticancer properties [25] Antimutagenic and immunomodulatory activities [23] Protective against lung cancer and improved lung function [28]
Lutein	Tomato, goji berry, romaine lettuce, zucchini, kiwifruit, garden peas, olive [19], parsley, broccoli, avocado, Brussels sprouts, beans [25], corn [21]	Protective action against ocular diseases such as macular degeneration and cataract [21] Antioxidant agents [23] improves visual acuity, scavenges harmful ROS [24]
Zeaxanthin	Same as lutein Mandarins, peaches, oranges [25]	Protects against macular degeneration and cataract [21] Antioxidant properties [23] improves visual acuity, scavenges harmful ROS [24]

Table 1.
Major dietary carotenoids and their health properties.

xanthophylls; β-cryptoxanthin, lutein, and zeaxanthin. These nutritionally important carotenoids are of major interest because they are detectable in the human plasma and can further be classified into provitamin A and non-provitamin A carotenoids. Provitamin A carotenoids are β-carotene, α-carotene, and β-cryptoxanthin, whereas non-provitamin A carotenoids are lutein, zeaxanthin, and lycopene [19]. Provitamin-A carotenoids are a major source of vitamin A, when ingested by human beings, they are converted into vitamin A, which has several important functions including vision, immune response, bone mineralization, reproduction, cell differentiation, and growth [20].

β-Carotene is the most widely distributed carotenoid in the human diet. α-Carotene is usually detected in similar foodstuff as β-carotene but in lower quantities. **Table 1** summarizes the main carotenoids in foodstuff and their effects on human health.

3. Carotenoids in tomato and processed products

3.1 Carotenoids in fresh tomato

Carotenoids are highly abundant in tomatoes [29]. Over 20 carotenoids have been previously characterized in tomato and tomato-based products, this includes lycopene, α-carotene, β-carotene, γ-carotene, ξ-carotene, ζ-carotene, phytoene, phytofluene, cyclolycopene, neurosporene, lutein, violaxanthin, neoxanthin, zeaxanthin, α-cryptoxanthin, and β-cryptoxanthin [26, 27]. The carotenoid content in tomato fruits is unevenly distributed and its composition is highly dependent on the cultivar (genotype), degree of maturation, climatic conditions, environmental factors, and cultural practices [7, 22, 30]. The maximum quantity of total carotenoids and lycopene is found in the outer pericarp, while the locules have a high proportion of carotene compounds [31].

Carotenoids are synthesized in the leaves, flowers, and fruits of tomato plants. Lutein is found in high quantities in the leaves where it functions as a photoreceptor during photosynthesis. The xanthophylls, violaxanthin, and neoxanthin are abundant

Tomato type	Analytical method	Carotenoid concentrations [mg/100 g fresh weight (FW)]	Reference
Cherry tomato	High-performance liquid chromatography with diode-array detection (HPLC-DAD)	Phytoene (0.43–2.01) Phytofluene (0.12–0.8) β-Carotene (1.16–4.15) Lycopene (0.17–9.66)	[32]
Industrial (processing) tomato	HPLC-DAD and HPLC	Phytoene (5.57–10.75) Phytofluene (1.89–3.55) ζ-Carotene (3.01–7.07) Neurosporene (0.8–1.74) β-Carotene (0.23–0.45) Lycopene (3.51–11.61) Lutein (0.076–0.429)	[32, 33]
Tomato for salad	HPLC	Lutein (0.077–0.338) Lycopene (5.18–8.47) β-Carotene (0.29–0.62)	[33]

Table 2.
Mean carotenoid composition of ripe fruits of different types of tomato.

in flowers and are responsible for their characteristic yellow coloration. The ripe fruits of *Lycopersicum esculentum* owe their intense red color to the lycopene, the main carotenoid at this maturity stage [7]. The most widely available carotenoids in ripe tomato fruits are lycopene (≈90%), β-carotene (≈5–10%), and lutein (<1%). Lycopene and β-carotene are mainly responsible for the characteristic color of ripe tomatoes [30]. These carotenoids are important components for the determination of quality characteristics of fresh tomatoes in routine analyses [16, 30]. **Table 2** shows the carotenoid profile of cultivars from different types of tomatoes.

There is a diverse carotenoid profile within tomato cultivars. This is particularly true for traditional varieties constituting a wide source of genetic variation [24]. Tomatoes are abundant sources of lycopene, with average concentrations ranging from 8 to 40 μg/100 g of FW. This represents about 80% of the total dietary intake of this carotenoid [34]. Lycopene is a polyunsaturated compound containing 13 double bonds that can exist in *trans*- and *cis* configurations. In fresh tomatoes, lycopene is principally found in *trans*-conformation [35]. Lycopene is synthesized massively during tomato ripening. Consequently, the highest content of lycopene is observed in ripe tomato fruits [7, 29, 27]. Open field-cultivated tomatoes were reported to have a higher lycopene content (ranging from 5.2 to 23.6 mg/100 g FW) than greenhouse-cultivated tomatoes (0.1 to 10.8 mg/100 g FW) [35].

The β-carotene content in tomatoes is approximately one-tenth of the lycopene content [31]. β-Carotene is equally an essential carotenoid identified in tomatoes, of special interest mainly due to its pro-vitamin A activity [33]. In commercial cherry tomatoes, β-carotene quantity reached 1.26 mg/100 g FW (**Table 2**). The uniqueness of β-carotene is that it is the most powerful precursor to vitamin A (comprised of retinol, retinal, and retinoic acid, which are classified as retinoids). Vitamin A activity can be measured as retinol equivalents (RE) or retinol activity equivalents (RAE). Current assumptions regarding the RAE or RE of the three major provitamin A dietary carotenoids based on their bioavailability from foods, consider β-carotene as a prominent contributor to the vitamin A intake with potential for conversion to retinol, which is twice that of α-carotene and β-cryptoxanthin [36]. The central oxidative cleavage of β-carotene in the intestine catalyzed by β-carotene 15,15′-monooxygenase allows for its conversion to two molecules of vitamin A, compared to one molecule from another provitamin A carotenoids [37]. Lesser amounts of lutein are present in tomatoes, with concentrations up to 338 μg/100 g FW (**Table 2**). Raw tomato purchased from the supermarket was reported to have lutein concentrations up to 32 μg/100 g FW, against a lutein content up to 800 μg/100 g FW reported in a cherry tomato variety [35]. Other carotenoids identified in tomatoes are the colorless hydrocarbon carotenoids (carotenes), phytoene, and phytofluene, precursors of colorful carotenoids such as lycopene and β-carotene [24].

3.2 Carotenoids in processed tomato products

Although tomatoes are consumed fresh, over 80% of tomato intake is in the form of processed products, such as tomato pulp, ketchup, juice, and sauce [38]. During food processing, the naturally occurring carotenoid composition of products is altered. Reactions induced by heat, acids, light, or oxygen exposure occur as a consequence of the processing steps [39]. Thermal treatment is responsible for an increased level of total carotenoid content and antioxidant capacity by 30% and 15%, respectively. Tomato processing may activate the enzymes ε- and β-carotene cyclase, involved in the synthesis of β- and α-carotene. Consequently, stimulating the

Tomato Product	Analytical method	Carotenoid concentrations (mg/100 g FW)	Reference
Ketchup	HPLC	Lycopene (18.80–100.87) β-Carotene (0.46–10)	[41]
Canned cherry tomatoes	HPLC	Lycopene (11.42–11.78) β-Carotene (0.74–0.76) Lutein (0.14–0.16)	[42]
Tomato Purée	HPLC	Lycopene (53.36–128.60) β-Carotene (0.40–2.80)	[41]

Table 3.
Concentration of carotenoids in processed tomato products.

production of α- and β-carotene [40]. The concentrations of carotenoids in different tomato products are depicted in **Table 3**.

During tomato processing, an increase in carotenoid content on a fresh weight basis is observed as a result of water loss [41]. This may also be ascribed to the technological treatments of pasteurization and homogenization which can improve the extractability of pigments from the fruit matrix. For canned tomato products, the carotenoid increase can be explained by the use of tomato juice derived from high ripening stage tomatoes with very high lycopene content [42]. Increased content of the major tomato carotenoids, lycopene, and β-carotene was reported after processing at 45°C (drying) and 95°C (thermal treatment of tomato juice) [43]. Similarly, an increase in lycopene content in tomatoes exposed to drying at 42°C was previously demonstrated. This occurs due to the release of lycopene bound from the tissues [44]. A decrease in the lycopene content of dried tomatoes treated at 55–110°C was found [45–47]. On a dry weight basis, there is an increase or decrease of the lycopene content depending on the origin of the tomato variety, while the β-carotene content reduces or remains relatively constant [41]. Nevertheless, in certain instances, processing causes little or no change in the content and activity of naturally occurring bioactive compounds [48].

4. Bioavailability of tomato carotenoids

Only 25 carotenoids are present in the human bloodstream, out of approximately 40 carotenoids present in foods normally included in the human diet and most of these carotenoids found in human blood are present just in fresh tomato and related products [43]. This is due to the selective intake of carotenoids in the gastrointestinal tract and the food matrix surrounding them [16, 43]. Carotenoids present in the human serum tend to be associated with specific body tissues. For example, lycopene is concentrated in the prostate, β-carotene is concentrated in the corpus luteum, and lutein and zeaxanthin are concentrated in the neural retina and brain neocortex. These carotenoids can retard the development of disease at these locations based on reducing inflammatory and oxidative stress [49]. For carotenoid intake, the food matrix made up of fiber or protein must first be broken down by mastication, gastric acid, pancreatic enzymes, and bile acids to ensure the release of these nutrients [16]. Carotenoid release from the tomato matrix and its subsequent incorporation in the oil and micellar phase are crucial steps in rendering these compounds bioavailable during digestion [20]. There is a great variation in the bioaccessibility and bioavailability of different dietary carotenoids between the type of food consumed (whether it is chopped or pureed, raw or cooked, and whether or not fat is consumed simultaneously), and for a given carotenoid in different foods [36, 50]. Bioaccessibility is defined as the fraction of carotenoid released during digestion from the food matrix to mixed micelles and thus, made accessible for absorption in the gut following digestion [51], whereas bioavailability of carotenoids is the amount of these micronutrients that are absorbed by the intestinal absorptive cell, transported in the bloodstream and/or deposited in target tissues where it can exert its biological function [52].

4.1 Processing effects on tomato carotenoid bioavailability

The bioavailability of carotenoids is higher from processed foods than their raw or less processed counterparts [52]. In general, the relative bioavailability of carotenoids has been estimated to vary from less than 10% in raw, uncooked vegetables to 50% in oils or commercial preparations [50]. Processing techniques such as grinding, marinating, fermentation, freezing, and moderate heating improve the release and absorption of carotenoids. This is explained by the release of these nutrients from the food matrix as a result of the disruption of plant tissues and the transfer of carotenoids to the lipid carrier. It is believed that since carotenoids in plant tissues occur in the form of complexes with proteins, mild thermal processes allow them to break down these connections and destroy cellulose structures in plant cells, thus contributing to an increase in the absorption of these compounds [53]. The bioavailability of β-carotene is improved as a result of gentle heating or enzymatic disruption of the vegetable cell wall structure during processing [48]. Lycopene bioavailability is higher in thermally processed tomato products, such as paste, puree, ketchup, juice, soup, and sauce, than in fresh tomatoes [33, 35, 54]. This fact could be attributed to the lower availability of lycopene from the raw tomatoes where it is probably bound in the surrounding food matrix [55]. The incorporation of oil in tomato sauce has been reported to enhance the accessibility and extractability of carotenoid compounds in tomatoes. A constant quantity of fat and other ingredients significantly increases the bioavailable lycopene in tomato paste compared to fresh tomatoes [40]. Previous research demonstrated that a combination of homogenization and heat treatment improves the bioavailability of carotenoids from fruits and vegetables. Studies on the effect of heat treatment and homogenization on the carotenoid bioavailability of industrially heat-treated peeled and canned tomatoes have shown that blood plasma lycopene responses increased with increasing degree of homogenization and additional heat treatment, while homogenization enhanced the plasma response of β-carotene only if the tomatoes were not subjected to additional heat treatment [56]. Moreover, high-pressure homogenization has a greater impact on the bioavailability of carotenoids compared to homogenization under normal pressure, since it disrupts extra cell membranes [42].

4.2 Effect of isomerization on bioavailability

The physical state of carotenoids has been proven to significantly impact their bioaccessibility and bioavailability and consequently their health-promoting properties [39]. Carotenoids exist in a variety of geometric isomers and predominantly occur in their all-trans conformation in fresh tomatoes. For instance, trans-lycopene accounts for approximately 95% of the lycopene present in raw tomatoes [48]. Food processing may induce the formation of cis isomers possessing different biological properties. Trans-to-cis isomerization can also be initiated during storage [55]. Trans-isomers are thermodynamically more stable, whereas cis are more polar, more soluble in oil and hydrocarbon solvents, and are less prone to crystallization than their all-trans counterparts [38, 55]. More than 50% of the carotenoids identified in the human body are in the cis configuration, suggesting that this is the most bioavailable form [40]. Several reports have demonstrated that the cis isomers of lycopene are more bioavailable and play a more important biological function than all-trans lycopene properties [57, 58] because of being more soluble and easily absorbed from the intestinal lumen than the trans-lycopene [59]. Therefore, lycopene from processed tomato products is generally more bioavailable than the one from the unprocessed counterparts. Nevertheless, inadequate processing and storage conditions can cause isomerization during the byproducts' formation, diminishing the absorption of carotenoids and making the product less desirable to the consumer [20]. On the other hand, cellular studies reported that cis isomers of β-carotene are not easily absorbed by intestinal enterocytes. High quantities of cis isomers of ß-carotene are not detected in the bloodstream, suggesting preferential absorption of the all-trans isomer of nutrients possessing provitamin A activity [52].

5. Role of tomato carotenoids in chronic disease prevention

Consumption of fruits and vegetables with beneficial health properties has been exploited for their ability to treat or prevent several chronic diseases [60]. There is an inverse relationship between the balanced consumption of tomatoes and tomato-derived products and the incidence of chronic diseases such as CVDs and various forms of cancers. These beneficial effects are attributed to carotenoids and phenolic compounds, which have high antioxidant capacities [48].

5.1 Oxidative stress

Oxidative stress plays an essential pathophysiological role in various chronic diseases such as CVDs, diabetes, neurodegenerative diseases, and cancer [60]. Free radicals, or other reactive oxygen- or nitrogen-containing species, are responsible for oxidative stress [48]. Oxidative stress occurs when there is a relative excess of ROS when compared with antioxidants [61]. ROS are reduced oxygen metabolites characterized by strong oxidizing capabilities. They are deleterious to cells at high concentrations but at low concentrations, they play a major role in cellular signaling and function [62]. ROS are formed as a by-product of mitochondrial respiration or metabolic activities (such as breathing, digesting food, metabolizing alcohol and drugs, and turning fats into energy) or by enzymes, such as superoxide dismutase, glutathione peroxidase, catalase, peroxiredoxins, and myeloperoxidases [60, 63]. Cells possess complex biochemical and genetic mechanisms to maintain ROS at physiologically normal concentrations, and deregulation in this balance has detrimental health effects [61, 62]. This is because abnormally high ROS levels may attack certain biomolecules (DNA, RNA, proteins lipids, and carbohydrates) causing damage to cells, tissues, and organs. [64]. The continuous production of free radicals in humans must be equivalent to the rate of antioxidant intake/synthesis [60]. Molecules such as ascorbate, a-tocopherol, and carotenoids are examples of antioxidants that are capable of quenching ROS. The structural properties of carotenoid molecules, particularly the presence of conjugated carbon–carbon double bonds enable the quenching of ROS and subsequently a reduction in ROS levels [19]. Tomato and related products contain carotenoids, particularly lycopene, one of the most potent antioxidants that have been found to protect against these chronic diseases by mitigating oxidative damage and improving the oxidative status [20, 48, 65]. Lycopene exerts strong antioxidant activity because it contains many double-conjugated bonds (11 conjugated double bonds and two unconjugated double bonds), which explains why lycopene can quench ROS and efficiently scavenge free radicals [7]. A study demonstrated that a long-term tomato-rich diet consisting of tomato juice, tomato sauce, tomato paste, ketchup, spaghetti sauce, and ready-to-serve tomato soup can reduce oxidative stress, this was attributed to an increase in serum lycopene levels from 181.79 ± 31.25 to 684.7 ± 113.91 nmol/l, as well as an increase in total antioxidant potential from 2.26 ± 0.015 to 2.38 ± 0.17 mmol/l Trolox equivalent [65]. The level of oxidative stress induced by in-vitro X-ray exposure in healthy adults was determined using serum 8-oxo-7, 8-dihydro-2-deoxyguanosine (8-oxo-dG), and plasma reactive oxygen metabolite-derived compounds (d-ROMs), the results suggested that continuous tomato juice consumption could decrease extracellular 8-oxo-dG and d-ROMs [66]. Previous studies have shown that tomato extracts containing 6% lycopene, other tomato carotenoids (phytoene and phytofluene above 1%, beta-carotene above 0.2%), can prevent oxidative stress-induced damage to fibroblast skin cells [67].

5.2 Tomato consumption and cardiovascular diseases

Worldwide, CVDs are an increasing concern due to the rising prevalence and consequent mortality and disability with a heavy economic burden since it is an important contributor to the cost of medical care [68, 69]. In 2019, 17.9 million people died from CVDs, representing 32% of all global deaths [70]. There is a growing body of epidemiological evidence that tomato and tomato products intake lower the risk of CVDs, through antioxidative, anti-inflammatory, and hypotensive effects [71]. The improvement of biomarkers associated with CVD development and the subsequent reduction in CVD risk has been ascribed to increased plasma lycopene levels. Moderate intake (2–4 servings) of tomato products such as soup, paste puree, juice, or any other tomato beverages, when consumed with the addition of dietary lipids, such as olive oil or avocados, leads to a rise in plasma carotenoids, particularly lycopene [72]. Dietary lycopene consumed as oil-based tomato products confers cardiovascular benefits. The consumption of ≥7 servings/week of tomato-based food products has been associated with a 30% reduction in CVD development in women [73]. Consumption of two glasses of tomato juice satisfies the recommended daily intake of lycopene (35 mg), [74]. **Table 4** shows the lycopene content of tomatoes and some frequently consumed tomato-derived products.

Epidemiological studies also suggest that the risk of myocardial infarction is lowered in individuals with higher lycopene content in adipose tissue. The EURAMIC (European community multicenter study on antioxidants, myocardial Infarction, and breast cancer) case–control study conducted in 10 European countries to assess the relations between antioxidant status and acute myocardial infarction, found lycopene concentration of adipose tissue to be independently protective against myocardial infarction [76]. A recent study by Cheng et al. [77] reported that higher intakes of lycopene or its high serum concentrate have been associated with significant reductions in the risk of stroke (26%) and CVDs (14%). Another carotenoid present in processed tomato products associated with CVD risk reduction is β-carotene [78].

Low levels of high-density lipoprotein (HDL) cholesterol and elevated LDL cholesterol are established CVD risk factors [79]. Pharmacological therapies aimed at LDL lowering have convincingly proven to reduce CVD disorders, such as coronary

Product	Lycopene (mg/100 g)
Fresh tomatoes	0.72–20
Tomato juice	5–11.60
Tomato puree	16.67–34.7
Tomato paste	5.40–150.00
Ketchup	9.90–17.00

Adapted from [72, 74, 75].

Table 4.
Lycopene content of tomatoes and processed tomato products.

heart disease. Therefore, LDL cholesterol levels should be lowered as much as possible to prevent CVD [80, 81]. Lycopene may modulate the expression of adhesion molecules in human vascular endothelial cells and increase the expression of LDL receptors involved in the regulation of cholesterol metabolism [75]. Increasing the concentration of HDL can slow and even reverse the progression of coronary atherosclerosis (coronary heart disease) and reduce CVD risk in those with dyslipidemia (abnormal levels of blood lipids including cholesterol). Consumption of two uncooked tomatoes per day demonstrated a significant elevation of HDL levels in overweight women [79]. Michaličková et al. [71] conducted a randomized controlled study to examine the effect of tomato juice on LDL cholesterol. The intervention group was supplemented with 200 g of tomato juice for 4 weeks and a significant reduction in total cholesterol and LDL was observed [71] indicating that tomato and derivatives have favorable effects on lipid metabolism.

Systemic arterial hypertension is a condition in which an individual has abnormally high blood pressure (BP) and is a primary risk factor for CVDs [82]. BP above 140 mmHg systolic and/or 90 mmHg diastolic is considered hypertensive [83]. Several studies indicated that tomato products intake leads to a significant reduction in BP [84, 85]. A higher dosage of tomato-derived supplements (containing more than 12 mg lycopene per day) could significantly lower systolic blood pressure (SBP), particularly among populations with baseline SBP > 120 mmHg [84]. The effect of treatments with tomato nutrient complexes (containing 5, 15, and 30 mg lycopene) was compared with 15 mg of synthetic lycopene and a placebo over 8 weeks, significant reductions in mean SBP were noted in tomato nutrient complexes treatments with 15 or 30 mg of lycopene [86]. A recent trial highlighted the benefits of processed tomato products on BP management in overweight middle-aged adults. A lowered diastolic BP was observed in participants that consumed a high tomato diet consisting of approximately 200 g/day or 1400 g/week of tomato products [78]. In a quasi-experimental study, 32 type 2 diabetes patients consumed 200 g raw tomato daily for 8 weeks. A significant decrease in systolic and diastolic BP was noted at the end of the study compared with initial values [87]. Tomato consumption might be beneficial for reducing CVD risk in type 2 diabetic patients.

5.3 Anti-cancer role of tomato carotenoids

The consumption of tomatoes and tomato-derived products is inversely related to the incidence of different types of cancers, (prostate, stomach, and lung cancers) [7, 72, 88, 89]. A study on elderly patients in the US attributed a 50% reduction in mortality rates from cancer of all sites to a high intake of tomatoes [90]. Tomatoes and tomato products are typical components of the Mediterranean diet (MD).

The MD represents a dietary pattern suitable for the prevention of chronic diseases [91]. A meta-analysis of observational studies, which evaluated the effects of the adoption of the MD on incidence and mortality of different types of cancer, showed that the high adherence to this diet was associated with a significantly lowered risk of overall cancer, especially colorectal cancer, pharyngeal and esophageal cancer, and prostate cancer [92]. The protective role of tomatoes is predominantly ascribed to the carotenoid, lycopene [93]. Researchers found that there was a lower rate of mortality from cancer in the group of US adults with the highest tomato and lycopene intake (42.5% and 45.9%, respectively) [94].

Extensive research has been conducted on the role of lycopene in the prevention of prostate cancer, the second most frequent cancer (after lung cancer) diagnosed in men worldwide [7, 72, 95], with higher incidence and mortality observed in developed countries [96]. Findings from ecological and migrant studies suggest that the wide disparity in incidence rates of prostate cancer worldwide may be attributed to a "Westernized" diet and lifestyle in developed countries [97]. A study conducted in 2011 using DU145 cells (human prostate cancer cells), revealed that the proliferation of these cells was significantly inhibited by lycopene. The authors found that lycopene induced a reduction of the proliferation rate at concentrations of 15 and 25 μM, but not at physiological concentrations (>2 μM) [98]. The US health Professionals Follow-up Study investigated the relationship of various carotenoids and retinol consumption with the risk of prostate cancer. There was an inverse relationship between the estimated intake of lycopene and the risk of this cancer. This reduced incidence was not observed with any other carotenoid. A reduction in risk of almost 35% was observed for a consumption rate of 10 or more servings of tomato products per week, and the protective action was greater with more advanced or aggressive prostate cancer [95]. In a more recent study, there was an 18% lower risk of prostate cancer associated with adherence to the same recommended tomato intake [97].

Evidence pointing to the protective effect of tomato product consumption for other cancer sites other than the prostate is ambiguous [99]. Lung cancer is the leading cause of cancer death, with an estimated 1.8 million deaths (18%) [100]. Growing evidence suggests that tomato lycopene may be preventive against the development of this cancer [101]. In 2020, a study demonstrated that lycopene treatment may inhibit the growth of lung tumor cell line A549. Varying amounts of lycopene (2.5, 5, and 25 μL) were used to treat lung cancer cell cultures and higher lycopene concentrations were more damaging to cancer cell nuclei [102]. Among 14 case-control lung cancer studies, only 6 studies showed a statistically significant risk reduction for cancer incidence, averaging 51%. However, cohort studies showed no beneficial relation between lung cancer reduction and tomato product consumption [99]. According to epidemiological studies, higher lycopene intake is associated with either a reduced or no change in lung cancer risk when compared to lower intake levels [103]. Gastric (stomach) cancer remains one of the dominant causes of cancer mortality in the world [104, 105]. Tomato or lycopene intake has proven to reduce gastric cancer risk in a variety of populations [72, 95]. However, few studies have been conducted to date. A meta-analysis study consisting of 21 studies supports an inverse association between tomato consumption and risk of gastric cancer [106]. Previous research projects have reported a negative relationship between tomato intake and the risk of gastric cancer. A study conducted in Korea consisting of 1245 subjects (415 cases and 830 matched controls; 810 men and 435 women), highlighted that the consumption of tomatoes and tomato ketchup was inversely associated with GC risk in the overall subjects [107]. In a case-control study in Uruguay, tomato consumption had a strong inverse association with gastric carcinogenesis. The carotenoids, α-carotene, and lycopene were strongly associated with this reduction in stomach cancer development [108].

5.4 Tomato in protection against obesity and diabetes

The incidence of type 2 diabetes (diabetes mellitus) and obesity has increased worldwide during the last century in both developed and developing countries [109]. Obesity is a chronic inflammatory disorder in which an increase in circulating inflammatory mediators is caused by an increase in body fat [20, 110]. Destructive mechanisms associated with obesity increase ROS and hamper the antioxidant status [111]. Individuals having a fasting blood sugar level of 126 mg/dl or higher on 2 separate days, will be diagnosed with type 2 diabetes [112]. The strong link between type 2 diabetes and obesity [113], with 80 percent of type 2 diabetes patients being overweight [112], was named "diabesity." According to the WHO, overweight and obesity account for 44% of diabetes cases. Therefore, it is necessary to develop therapeutic strategies favoring weight loss and blood glucose control (anti-obesity and antidiabetic treatment) [114].

A randomized controlled clinical trial was conducted on 64 overweight or obese demonstrated that tomato juice reduces oxidative stress in overweight females and may prevent the development of obesity-related diseases. In this study, the antioxidant parameters of study participants that ingested 330 ml/day of tomato juice for 20 days were analyzed at the beginning and after this period verifying an increase in plasma total antioxidant capacity (TAC) and erythrocyte antioxidant enzymes [115]. Ghavipour et al. [110] demonstrated that tomato juice consumption lowers inflammation in overweight and obese females. The predictive biomarkers of inflammation [tumour necrosis factor-alpha (TNF-α) and interleukin 8 (IL-8)] were examined in study participants who drank 330 ml of tomato juice every day for 20 days. The serum levels of IL-8 and TNF- α were significantly lower in overweight people that consumed the tomato juice compared to the control group. The scientists concluded that eating more tomatoes may lower the risk of inflammatory disorders, such as CVDs and diabetes [110].

The goal of diabetes management is to maintain plasma glucose concentrations at near-normal levels [112]. According to the WHO, expected values for normal fasting blood glucose levels are between 70 mg/dl (3.9 mmol/l) and 100 mg/dl (5.6 mmol/l) [116]. Chemicals found in fresh or processed tomatoes have been shown to have antihyperglycemic properties that enable the lowering of glucose levels in the blood. In streptozotocin (STZ)-induced hyperglycemic rats, oral administration of tomato extract lycopene (90 mg/kg of body weight) resulted in a lower serum glucose level. The therapeutic amount of lycopene in humans is around 14.5 mg/kg of body weight. Lycopene's anti-diabetic properties may be linked to its antioxidant activity, which reduces the number of free radicals generated [117]. Another study indicated that fasting blood sugar levels decreased after drinking tomato juice for 3 weeks [112]. The reduction in fasting blood glucose levels was found to be an average of 9.00 mg/dl (7.64%). Supplementation with β-carotene did not affect type 2 diabetes in randomized controlled trials [118, 119]. The impact of lycopene consumption on blood glucose concentration was analyzed, each 1 mg increase in lycopene consumption was associated with a 0·005 mmol/l decrease in fasting blood glucose concentration [120]. The effects of pre-prandial tomato intake on body weight, fat percentage, triglyceride, cholesterol, and blood sugar levels were evaluated in 35 young women aged 18 to 21 years. Participants ate raw, ripe tomatoes (90 g) before lunch each day for 4 weeks. At the end of the study, there were significant reductions in body weight (1.09 ± 0.12 kg), fat % (1.54 ± 0.52%), fasting blood glucose (5.29 ± 0.80 mg/dl), triglycerides (8.31 ± 1.34 mg), and cholesterol (10.17 ± 1.21 mg/dl). Thus, tomato consumption before meals was positively correlated with body weight, fat %, triglycerides, blood sugar, and cholesterol levels in young adult women [121].

6. Conclusion

Tomato is a food product available all year round and is highly consumed by populations around the world. Tomato carotenoids have demonstrated antioxidant and protective effects against chronic diseases. Among these carotenoids, lycopene, in particular, has shown distinct antioxidant and anticancer properties at cellular levels. Numerous studies highlighted the potential benefits of tomato carotenoids in delaying or preventing the development of chronic degenerative diseases. Nevertheless, further research is required to better elucidate the beneficial health effects of these carotenoids as well as their precise modes of action in the risk reduction of chronic diseases. Considering the reported positive implications of tomatoes and their products in chronic disease prevention, dietary intake of naturally occurring carotenoid-rich tomato and processed tomato products should be highlighted and recommended.

Acknowledgements

This work was financed by the ongoing project PRIMA H2020 GA2032, FunTomP—Functionalized Tomato Products (https://funtomp.com/), a multidisciplinary project involving 16 countries, that aims to reformulate traditional Mediterranean tomato products into different functional foods using leaf proteins (by-products of sugar beet processing) and olive powder and novel and eco-friendly processing technologies that will minimally affect nutrients, with extra health benefits while keeping a sustainable product and process cycle and by valorizing agricultural waste.

Author details

Rose Daphnee Ngameni Tchonkouang[1], Maria Dulce Carlos Antunes[2] and Maria Margarida Cortês Vieira[1*]

1 Department of Food Engineering, University of Algarve Faro, Portugal

2 Mediterranean Institute for Agriculture, Environment and Development, University of Algarve Faro, Portugal

*Address all correspondence to: mvieira@ualg.pt

References

[1] Bergougnoux V. The history of tomato: From domestication to biopharming. Biotechnology Advances. 2014;**32**(1):170-189. DOI: 10.1016/j.biotechadv.2013.11.003

[2] Kimura S, Sinha N. Tomato (*Solanum lycopersicum*): A model fruit-bearing crop. Cold Spring Harbor Protocols. 2008;**3**:1-9. DOI: 10.1101/pdb.emo105

[3] Villanueva EE. An Overview of Recent Studies of Tomato (*Solanum lycopersicum* spp.) from a Social, Biochemical and Genetic Perspective on Quality Parameters [Introductory Paper]. Alnarp-Sweden: Swedish University of Agricultural Sciences; 2018

[4] Noonari S, Solangi SU, Laghari MA, Wagan SA, Sethar AA, Kalwar GY, et al. Economic implications of tomato production in Naushahro Feroze District of Sindh Pakistan. Research on Humanities and Social Sciences. 2015;**5**:158-170

[5] Quinet M, Angosto T, Yuste-Lisbona FJ, Blanchard-Gros R, Bigot S, Martinez JP, et al. Tomato fruit development and metabolism. Frontiers in Plant Science. 2019;**10**(1554):1-23. DOI: 10.3389/fpls.2019.01554

[6] Food and Agriculture Organization (FAO). FAOSTAT: Crops and Livestock Products [Internet]. 2022. Available from: https://www.fao.org/faostat/en/#data/QCL/visualize [Accessed: 26 January 2022]

[7] Martí R, Roselló S, Cebolla-Cornejo J. Tomato as a source of carotenoids and polyphenols targeted to cancer prevention. Cancers. 2016;**8**(58):2-28. DOI: 10.3390/cancers8060058

[8] Stajčić S, Ćetković G, Čanadanović-Brunet J, Djilas SM, Mandic AI, Četojević-Simin DD. Tomato waste: Carotenoids content, antioxidant and cell growth activities. Food Chemistry. 2015;**172**:225-232. DOI: 10.1016/j.foodchem.2014.09.069

[9] Carotene.org. Carotenoids – An Overview [Internet]. 2022. Available from: https://www.carotene.org/carotenoids/ [Accessed: 26 January 2022]

[10] Britton G, Liaaen-Jensen S, Pfander H, editors. Carotenoids Handbook. 1st ed. Basel: Birkhauser Verlag; 2004. p. 885. DOI: 10.1007/978-3-0348-7836-4

[11] Ruiz-Sola MÁ, Rodríguez-Concepción M. Carotenoid biosynthesis in Arabidopsis: A colorful pathway. Arabidopsis Book. 2012;**10**:1-28. DOI: 10.1199/tab. 0158

[12] Swapnil P, Meena M, Singh SK, Dhuldhaj UP, Marwal A. Vital roles of carotenoids in plants and humans to deteriorate stress with its structure, biosynthesis, metabolic engineering and functional aspects. Current Plant Biology. 2021;**26**:1-11. DOI: 10.1016/j.cpb.2021.100203

[13] Maoka T. Carotenoids as natural functional pigments. Journal of Natural Medicines. 2020;**74**:1-16. DOI: 10.1007/s11418-019-01364-x

[14] Delgado-Vargas F, Jiménez AR, Paredes-López O. Natural pigments: Carotenoids, anthocyanins, and betalains - characteristics, biosynthesis, processing, and stability. Critical Reviews in Food Science and Nutrition. 2000;**40**(3):173-289. DOI: 10.1080/10408690091189257

[15] Maoka T. Carotenoids in marine animals. Marine Drugs. 2011;**9**(2):278-293. DOI: 10.3390/md9020278

[16] Eldahshan OA, Singab ANB. Carotenoids. Journal of Pharmacognosy and Phytochemistry. 2013;2(1):225-234

[17] Prakash D, Gupta C. Carotenoids: Chemistry and health benefits. In: Prakash D, Sharma G, editors. Phytochemicals of Nutraceutical Importance. 1st ed. Wallingford: CAB International publishing; 2014. pp. 181-195. DOI: 10.1079/9781780643632.0181

[18] Thomas SE, Johnson EJ. Xanthophylls. Advances in Nutrition. 2018;**9**(2):160-162. DOI: 10.1093/advances/nmx005

[19] Bakan E, Akbulut ZT, Inanç AL. Carotenoids in foods and their effects on human health. Akademik Gıda. 2014;**12**(2):61-68

[20] Pinela J, Oliveira MBPP, Ferreira ICFR. Bioactive compounds of tomatoes as health promoters. In: Silva LR, Silva BM, editors. Natural Bioactive Compounds from Fruits and Vegetables. 2nd ed. Sharjah: Bentham Science Publishers; 2016. pp. 48-91. DOI: 10.2174/97816810824311160101

[21] Rodriguez-Amaya DB, Kimura M. HarvestPlus Handbook for Carotenoid Analysis. 1st ed. Washington, DC: International Food Policy Research Institute; 2004. p. 63

[22] Elvira-Torales LI, García-Alonso J, Periago-Castón MJ. Nutritional importance of carotenoids and their effect on liver health: A review. Antioxidants (Basel). 2019;**8**(7):229-252. DOI: 10.3390/antiox8070229

[23] Sartori AGO, Silva MV. Main food sources of carotenoids, according to the purpose and degree of processing, for beneficiaries of the 'Bolsa Família' in Brazil. Food Science and Technology. 2014;**34**(2):408-415. DOI: 10.1590/fst.2014.0056

[24] Flores P, Sánchez E, et al. Genotypic variability of carotenoids in traditional tomato cultivars. Food Research International. 2017;**100**(Pt 3):510-516. DOI: 10.1016/j.foodres.2016.07.014

[25] Jaswir I, Noviendri D, Hasrini RF, et al. Carotenoids: Sources, medicinal properties and their application in food and nutraceutical industry. Journal of Medicinal Plants Research. 2011;**5**:7119-7131. DOI: 10.5897/JMPRX11.011

[26] Shi J, Kakuda Y, Yeung D. Antioxidative properties of lycopene and other carotenoids from tomatoes: Synergistic effects. BioFactors. 2004;**21**(1-4):203-210. DOI: 10.1002/biof.552210141

[27] Guil-Guerrero JL, Rebolloso-Fuentes MM. Nutrient composition and antioxidant activity of eight tomato (*Lycopersicon esculentum*) varieties. Journal of Food Composition and Analysis. 2009;**22**:123-129. DOI: 10.1016/j.jfca.2008.10.012

[28] Tanumihardjo SA. Carotenoids: Health effects. In: Caballero B, editor. Encyclopedia of Human Nutrition. 3rd ed. Cambridge: Academic Press; 2013. p. 292-297. DOI: 10.1016/B978-0-12-375083-9.00045-3

[29] Orsi B, Sestari I, Preczenhak AP, et al. Allelic variations in the tomato carotenoid pathway lead to pleiotropic effects on fruit ripening and nutritional quality. Postharvest Biology and Technology. 2021;181(111632):1-11. DOI: 10.1016/j.postharvbio.2021.111632

[30] Kotikova Z, Hejtmánková A, Jaromir L. Determination of the influence of variety and level of maturity on the content and development of carotenoids in tomatoes. Czech Journal of Food Sciences. 2009;**27**:200-203. DOI: 10.17221/1093-CJFS

[31] Inbaraj B, Bing-Huei C. Carotenoids in tomato plants. In: Preedy VR, Watson RR,

editors. Tomatoes and Tomato Products. 1st ed. Boca Raton: CRC Press; 2008. pp. 133-164. DOI: 10.1201/9781439843390-7

[32] Dzakovich MP, Gas-Pascual E, Orchard CJ, et al. Analysis of Tomato Carotenoids: Comparing Extraction and Chromatographic Methods. Journal of AOAC International. 2019;102:1069-1079. DOI: 10.5740/jaoacint.19-0017

[33] Abushita AA, Daood HG, Biacs PA. Change in carotenoids and antioxidant vitamins in tomato as a function of varietal and technological factors. Journal of Agricultural and Food Chemistry. 2000;**48**(6):2075-2081. DOI: 10.1021/jf990715p

[34] Del Giudice R, Petruk G, Raiola A, et al. Carotenoids in fresh and processed tomato (*Solanum lycopersicum*) fruits protect cells from oxidative stress injury. Journal of the Science of Food and Agriculture. 2016;**97**(5):1616-1623. DOI: 10.1002/jsfa.7910

[35] Raiola A, Rigano MM, Calafiore R, Frusciante L, Barone A. Enhancing the health-promoting effects of tomato fruit for biofortified food. Mediators of Inflammation. 2014;**2014**:1-16. DOI: 10.1155/2014/139873

[36] Olmedilla-Alonso B, Rodríguez-Rodríguez E, et al. Dietary β-cryptoxanthin and α-carotene have greater apparent bioavailability than β-carotene in subjects from countries with different dietary patterns. Nutrients. 2020;**12**(9):2639. DOI: 10.3390/nu12092639

[37] Nagarajan J, Ramanan RN, Raghunandan ME, et al. Carotenoids. In: Galanakis CM, editor. Nutraceutical and Functional Food Components: Effects of Innovative Processing Techniques. 1st ed. Amsterdam: Elsevier Inc; 2017. pp. 259-296. DOI: 10.1016/B978-0-12-805257-0.00008-9

[38] Gama J, Tadiotti A, Sylos C. Comparison of carotenoid content in tomato, tomato pulp and ketchup by liquid chromatography. Alimentos e Nutrição. 2009;**17**(4):353-358

[39] Schieber A, Weber F. Carotenoids. In: Carle R, Schweiggert R, editors. Handbook on Natural Pigments in Food and Beverages. Industrial Applications for Improving Food Color. 1st ed. Sawston: Woodhead Publishing; 2016. pp. 101-123. DOI: 10.1016/B978-0-08-100371-8.00005-1

[40] Vallverdú-Queralt A, Regueiro J, et al. Carotenoid profile of tomato sauces: Effect of cooking time and content of extra virgin olive oil. International Journal of Molecular Sciences. 2015;**16**(5):9588-9599. DOI: 10.3390/ijms16059588

[41] Schulz H, Schütze W, Baranska M. Fast determination of carotenoids in tomatoes and tomato products by Raman spectroscopy. Acta Horticulturae. 2006;**712**:901-905. DOI: 10.17660/ActaHortic.2006.712.118

[42] D'Evoli L, Lombardi-Boccia G, Lucarini M. Influence of heat treatments on carotenoid content of cherry tomatoes. Food. 2013;**2**(3):352-363. DOI: 10.3390/foods2030352

[43] Mendelová A, Fikselová M, Mendel Ľ. Carotenoids and lycopene content in fresh and dried tomato fruits and tomato juice. Acta Universitatis Agriculturae Et Silviculturae Mendelianae Brunensis. 2013;**61**(5):1329-1337. DOI: 10.11118/actaun201361051329

[44] Kerkhofs NS, Lister CE, Savage GP. Change in colour and antioxidant content of tomato cultivars following forced-air drying. Plant Foods for Human Nutrition. 2005;**60**:117-121. DOI: 10.1007/s11130-005-6839-8

[45] Zanoni B, Peri C, Nani R, Lavelli V. Oxidative heat damage of tomato halves as affected by drying. Food Research International. 1999;**31**(5):395-394. DOI: 10.1016/s0963-9969(98)00102-1

[46] Shi J, Le Maguer M, Kakuda Y, et al. Lycopene degradation and isomerisation in tomato dehydration. Food Research International. 1999;**32**(1):15-21. DOI: 10.1016/S0963-9969(99)00059-9

[47] Lavelli V, Hippeli S, Peri C, et al. Evaluation of radical scavenging activity of fresh and air-dried tomatoes by three model reactions. Journal of Agricultural and Food Chemistry. 1999;**47**(9):3826-3831. DOI: 10.1021/jf981372i

[48] Szabo K, Catoi AF, Vodnar DC. Bioactive compounds extracted from tomato processing by-products as a source of valuable nutrients. Plant Foods for Human Nutrition. 2018;**73**(4):268-277. DOI: 10.1007/s11130-018-0691-0

[49] Hammond BR, Renzi LM. Carotenoids. Advances in Nutrition. 2013;**4**(4):474-476. DOI: 10.3945/an.113.004028

[50] Deming D, Erdman J. Mammalian carotenoid absorption and metabolism. Pure and Applied Chemistry. 1999;**71**(12):2213-2223. DOI: 10.1351/pac199971122213

[51] Brown MJ, Ferruzzi MG, Nguyen ML, et al. Carotenoid bioavailability is higher from salads ingested with full-fat than with fat-reduced salad dressings as measured with electrochemical detection. The American Journal of Clinical Nutrition. 2004;**80**(2):396-403. DOI: 10.1093/ajcn/80.2.396 PMID: 15277161

[52] Kopec RE. Bioavailability, Metabolism, and Bioefficacy of Tomato Carotenoids [Thesis]. Columbus: Ohio State University; 2012

[53] Górecka D, Wawrzyniak A, et al. Lycopene in tomatoes and tomato products. Open Chemistry. 2020;**18**(1):752-756. DOI: 10.1515/chem-2020-0050

[54] Saad AG, Pék Z, et al. Determination of carotenoids in tomato products using Vis/Nir spectroscopy. The Journal of Microbiology, Biotechnology and Food Sciences. 2017;**7**(1):27-31. DOI: 10.15414/jmbfs.2017.7.1.27-31

[55] Shi J, Le Maguer M. Lycopene in tomatoes: Chemical and physical properties affected by food processing, Critical Reviews in Food Science and Nutrition. 2000;**40**(19):1-42, DOI: 10.1080/10408690091189275

[56] van het Hof KH, de Boer BC, Tijburg LB, et al. Carotenoid bioavailability in humans from tomatoes processed in different ways determined from the carotenoid response in the triglyceride-rich lipoprotein fraction of plasma after a single consumption and in plasma after four days of consumption. The Journal of Nutrition. 2000;**130**(5):1189-1196. DOI: 10.1093/jn/130.5.1189

[57] Lin CH, Chen BH. Determination of carotenoids in tomato juice by liquid chromatography. Journal of Chromatography A. 2003;**1012**(1):103-109. DOI: 10.1016/S0021-9673(03)01138-5

[58] Lin CH, Chen BH. Stability of carotenoids in tomato juice during storage. European Food Research and Technology. 2005;**221**:274-280. DOI: 10.1007/s00217-005-1155-y

[59] Strati I, Vasso O. Recovery and isomerization of carotenoids from tomato processing by-products. Waste and Biomass Valorization. 2016;**7**:843-850. DOI: 10.1007/s12649-016-9535-z

[60] Sharifi-Rad M, Kumar NVA, et al. Lifestyle, oxidative stress, and

antioxidants: Back and forth in the pathophysiology of chronic diseases. Frontiers in Physiology. 2020;**11**(649):1-21. DOI: 10.3389/fphys.2020.00694

[61] Hayes JD, Dinkova-Kostova AT, Tew KD. Oxidative stress in cancer. Cancer Cell. 2020;**38**(2):167-197. DOI: 10.1016/j.ccell.2020.06.001

[62] Arfin S, Jha NK, Jha SK, et al. Oxidative stress in cancer cell metabolism. Antioxidants (Basel). 2021;**10**(642):1-28. DOI: 10.3390/antiox10050642

[63] Dubois-Deruy E, Peugnet V, et al. Oxidative stress in cardiovascular diseases. Antioxidants. 2020;**9**(864):1-15. DOI: 10.3390/antiox9090864

[64] Engwa GA, EnNwekegwa FN, et al. Free radicals, oxidative stress-related diseases and antioxidant supplementation. Alternative Therapies in Health and Medicine. 2020;AT6236

[65] Rao AV. Processed tomato products as a source of dietary lycopene: Bioavailability and antioxidant properties. Canadian Journal of Dietetic Practice and Research. 2004;**65**(4): 161-165. DOI: 10.3148/65.4.2004.161

[66] Nakamura A, Itaki C, Saito A, et al. Possible benefits of tomato juice consumption: A pilot study on irradiated human lymphocytes from healthy donors. Nutrition Journal. 2017;**16**(27): 1-11. DOI: 10.1186/s12937-017-0248-3

[67] Darawsha A, Trachtenberg A, Levy J, et al. The protective effect of carotenoids, polyphenols, and Estradiol on dermal fibroblasts under oxidative stress. Antioxidants. 2021;**10**(12):2023. DOI: 10.3390/antiox10122023

[68] Amini M, Zayeri F, Salehi M. Trend analysis of cardiovascular disease mortality, incidence, and mortality-to-incidence ratio: Results from global burden of disease study 2017. BMC Public Health. 2021;**21**(401):1-12. DOI: 10.1186/s12889-021-10429-0

[69] Roth GA, Mensah GA, Johnson CO, et al. global burden of cardiovascular diseases and risk factors, 1990-2019: Update from the GBD 2019 study. Journal of the American College of Cardiology. 2020;**76**(25):2982-3021. DOI: 10.1016/j.jacc.2020.11.010

[70] World Health Organization (WHO). Cardiovascular Diseases (CVDs) [Internet]. 2022. Available from: https://www.who.int/news-room/fact-sheets/detail/cardiovascular-diseases-(cvds) [Accessed: 29 January 2022]

[71] Michaličková D, Belović M, Ilić N, et al. Comparison of polyphenol-enriched tomato juice and standard tomato juice for cardiovascular benefits in subjects with stage 1 hypertension: A randomized controlled study. Plant Foods for Human Nutrition;**74**:122-127. DOI: 10.1007/s11130-019-0714-5

[72] Marques C, Lima M, Oliveira J, et al. Tomato lycopene: Functional proprieties and health benefits. World Academy of Science, Engineering and Technology, International Journal of Agricultural and Biosystems Engineering. 2015;**9**(10): 1089-1099. DOI: 10.5281/zenodo.1109297

[73] Sesso HD, Liu S, Gaziano JM, Buring JE. Dietary lycopene, tomato-based food products and cardiovascular disease in women. The Journal of Nutrition. 2003;**133**(7):2336-2341. DOI: 10.1093/jn/133.7.2336

[74] Basuny MAM. The anti-atherogenic effects of lycopene. In: Frank S, Kostner G, editors. Lipoproteins - Role in Health and Diseases. London: IntechOpen; 2012. pp. 489-506. DOI: 10.5772/48134

[75] Thies F, Masson LF, Rudd A, et al. Effect of a tomato-rich diet on markers of cardiovascular disease risk in moderately overweight, disease-free, middle-aged adults: A randomized controlled trial. The American Journal of Clinical Nutrition 2012;95(5):1013-1022. DOI: 10.3945/ajcn.111.026286

[76] Willcox JK, Catignani GL, Lazarus S. Tomatoes and cardiovascular health. Critical Reviews in Food Science and Nutrition. 2003;**43**(1):1-18. DOI: 10.1080/10408690390826437

[77] Cheng HM, Koutsidis G, Lodge JK, et al. Lycopene and tomato and risk of cardiovascular diseases: A systematic review and meta-analysis of epidemiological evidence. Critical Reviews in Food Science and Nutrition. 2019;**59**(1):141-158. DOI: 10.1080/10408398.2017.1362630

[78] Burton-Freeman B, Edirisinghe I, et al. Processed tomato products and risk factors for cardiovascular disease. Nutrition and Aging. 2015;**3**(2-4):193-201. DOI: 10.3233/NUA-150059

[79] Cuevas-Ramos D, Almeda-Valdés P, et al. Effect of tomato consumption on high-density lipoprotein cholesterol level: A randomized, single-blinded, controlled clinical trial. Diabetes, Metabolic Syndrome and Obesity: Targets and Therapy. 2013;**6**:263-273. DOI: 10.2147/DMSO.S48858

[80] Rizvi AA, Stoian AP, Janez A, Rizzo M. Lipoproteins and cardiovascular disease: An update on the clinical significance of atherogenic small, dense LDL and new therapeutical options. Biomedicine. 2021;**9**(1579):1-7. DOI: 10.3390/biomedicines9111579

[81] Bantwal G, George B. Low density lipoprotein cholesterol target: Changing goal posts. Indian Journal of Endocrinology and Metabolism. 2012;**16**(Suppl. 2):233-235. DOI: 10.4103/2230-8210.104047

[82] Oparil S, Acelajado MC, Bakris GL, et al. Hypertension. Nature Reviews Disease Primers. 2018;**4**(18014):1-53. DOI: 10.1038/nrdp.2018.14

[83] Tan JL, Thakur K. Hypertension, Systolic. 1st ed. Treasure Island: StatPearls Publishing; 2018. p. 11

[84] Li X, Xu J. Lycopene supplement and blood pressure: An updated meta-analysis of intervention trials. Nutrients. 2013;**5**(9):3696-3712. DOI: 10.3390/nu5093696

[85] Ried K, Fakler P. Protective effect of lycopene on serum cholesterol and blood pressure: Meta-analyses of intervention trials. Maturitas. 2011;**68**(4):299-310. DOI: 10.1016/j.maturitas.2010.11.018

[86] Wolak T, Sharoni Y, Levy J, et al. Effect of tomato nutrient complex on blood pressure: A double blind. Randomized Dose-Response Study. Nutrients. 2019;**11**(950):1-13. DOI: 10.3390/nu11050950

[87] Shidfar F, Froghifa N, Vafa M, et al. The effects of tomato consumption on serum glucose, apolipoprotein B, apolipoprotein A-I, homocysteine and blood pressure in type 2 diabetic patients. International Journal of Food Sciences and Nutrition. 2011;**62**(3):289-294. DOI: 10.3109/09637486.2010.529072

[88] Giovannucci E, Ascherio A, Rimm EB, et al. Intake of carotenoids and retinol in relation to risk of prostate cancer. Journal of the National Cancer Institute. 1995;**87**(23):1767-1776. DOI: 10.1093/jnci/87.23.1767

[89] Rawla P. Epidemiology of prostate cancer. World Journal of Oncology.

2019;**10**(2):63-89. DOI: 10.14740/wjon1191

[90] Colditz GA, Branch LG, Lipnick RJ, et al. Increased green and yellow vegetable intake and lowered cancer deaths in an elderly population. The American Journal of Clinical Nutrition. 1985;**41**:32-36. DOI: 10.1093/ajcn/41.1.32

[91] Sofi F, Fabbri A, Casini A. Inflammation and Cardiovascular Disease and Protection by the Mediterranean Diet. In: Romagnolo DF, Selmin OI, editors. Mediterranean Diet. Dietary Guidelines and Impact on Health and Disease. 1st ed. Switzerland: Springer International Publishing; 2016. pp. 89-96

[92] Schwingshackl L, Hoffmann G. Adherence to Mediterranean diet and risk of cancer: A systematic review and meta-analysis of observational studies. International Journal of Cancer. 2014;**135**:1884-1897. DOI: 10.1002/ijc.28824

[93] Denniss SG, Haffner TD, Kroetsch JT, et al. Effect of short-term lycopene supplementation and postprandial dyslipidemia on plasma antioxidants and biomarker of endothelial health in young healthy individuals. Vascular Health and Risk Management. 2008;**4**:213-222. DOI: 10.2147/vhrm.2008.04.01.213

[94] Mazidi M, Ferns GA, Banach M. A high consumption of tomato and lycopene is associated with a lower risk of cancer mortality: Results from a multi-ethnic cohort. Public Health Nutrition. 2020;**23**(9):1569-1575. DOI: 10.1017/S1368980019003227

[95] Giovannucci E. Tomatoes, tomato-based products, lycopene, and cancer: Review of the epidemiologic literature. Journal of the National Cancer Institute. 1999;**91**(4):317-331. DOI: 10.1093/jnci/91.4.317

[96] Ferlay J, Shin HR, Bray F, et al. Estimates of worldwide burden of cancer in 2008: GLOBOCAN 2008. International Journal of Cancer. 2010;**127**(12):2893-2917. DOI: 10.1002/ijc.25516

[97] Hassanipour-Azgomi S, Mohammadian-Hafshejani A, Ghoncheh M, Towhidi F, Jamehshorani S, Salehiniya H. Incidence and mortality of prostate cancer and their relationship with the Human Development Index worldwide. Prostate International. 2016;**4**(3):118-124. DOI: 10.1016/j.prnil.2016.07.001

[98] Ford NA, Elsen AC, Zuniga K, et al. Lycopene and apo-12′-lycopenal reduce cell proliferation and alter cell cycle progression in human prostate cancer cells. Nutrition and Cancer. 2011;**63**(2):256-263. DOI: 10.1080/01635581.2011.523494

[99] Hwang ES, Bowen PE. Can the consumption of tomatoes or lycopene reduce cancer risk? Integrative Cancer Therapies. 2002;**1**(2):121-132. DOI: 10.1177/1534735402001002003

[100] Sung H, Ferlay J, Siegel RL, et al. Global cancer statistics 2020: GLOBOCAN estimates of incidence and mortality worldwide for 36 cancers in 185 countries. CA: A Cancer Journal for Clinicians. 2021;**71**(3):209-249. DOI: 10.3322/caac.21660

[101] Palozza P, Simone RE, et al. Tomato lycopene and lung cancer prevention: From experimental to human studies. Cancers. 2011;**3**(2):2333-2357. DOI: 10.3390/cancers3022333

[102] Pogoda K, Pucka M, Tabarkiewicz J, et al. Lycopene activity on lung and kidney cancer cells by T2 relaxation time 1H magnetic resonance imaging in vitro. European Journal of Clinical and Experimental Medicine. 2020;**18**(1):5-9. DOI: 10.15584/ejcem.2020.1.1

[103] Bacanli M, Başaran N, Başaran AA. Lycopene: Is it beneficial to human health as an antioxidant? Turkish Journal Of Pharmaceutical Sciences. 2017;**14**(3): 311-318. DOI: 10.4274/tjps.43043

[104] Crew KD, Neugut AI. Epidemiology of gastric cancer. World Journal of Gastroenterology. 2006;**12**(3):354-362. DOI: 10.3748/wjg.v12.i3.354

[105] Tsugane S, Sasazuki S. Diet and the risk of gastric cancer: Review of epidemiological evidence. Gastric Cancer. 2007;**10**(2):75-83. DOI: 10.1007/s10120-007-0420-0

[106] Yang T, Yang X, et al. The role of tomato products and lycopene in the prevention of gastric cancer: A meta-analysis of epidemiologic studies. Medical Hypotheses. 2013;**80**:383-388. DOI: 10.1016/j.mehy.2013.01.005

[107] Kim JH, Lee J, Choi IJ, et al. Dietary carotenoids intake and the risk of gastric cancer: A case-control study in Korea. Nutrients. 2018;**10**(8):1031. DOI: 10.3390/nu10081031

[108] De Stefani E, Boffetta P, Brennan P, et al. Dietary carotenoids and risk of gastric cancer: A case-control study in Uruguay. European Journal of Cancer Prevention. 2000;**9**(5):329-334. DOI: 10.1097/00008469-200010000-00007

[109] Serván PR. Obesity and diabetes. Nutricion Hospitalaria. 2013; **28**(Suppl. 5):138-143. DOI: 10.3305/nh.2013.28.sup5.6929

[110] Ghavipour M, Saedisomeolia A, Djalali M, et al. Tomato juice consumption reduces systemic inflammation in overweight and obese females. British Journal of Nutrition. 2013;**109**(11):2031-2035. DOI: 10.1017/S0007114512004278

[111] Pourahmadi Z, Mahboob S, Saedisomeolia A, et al. The effect of tomato juice consumption on antioxidant status in overweight and obese females. Women & Health. 2015;**55**(7):795-804. DOI: 10.1080/03630242.2015.1050546

[112] Tarigan H, Marlina S, et al. The Impact of Tomato Juice on Reducing Blood Sugar Levels at Diabetes Mellitus Patients, Proceedings of the International Conference on Health Informatics and Medical Application Technology (ICHIMAT 2019); 6-8 November 2019. Medan. Setúbal: SCITEPRESS; 2020. pp. 270-277

[113] Fagot-Campagna A, Balkau B, Simon D, et al. High free fatty acid concentration: An independent risk factor for hypertension in the Paris prospective study. International Journal of Epidemiology. 1998;**27**(5):808-813. DOI: 10.1093/ije/27.5.808

[114] Leitner DR, Frühbeck G, Yumuk V, et al. Obesity and type 2 diabetes: Two diseases with a need for combined treatment strategies - EASO can lead the way. Obesity Facts. 2017;**10**(5):483-492. DOI: 10.1159/000480525

[115] Ghavipour M, Sotoudeh G, Ghorbani M. Tomato juice consumption improves blood antioxidative biomarkers in overweight and obese females. Clinical Nutrition. 2015;**34**(5):805-809. DOI: 10.1016/j.clnu.2014.10.012

[116] World Health Organization. Mean fasting blood glucose [Internet]. 2022. Available from: https://www.who.int/data/gho/indicator-metadata-registry/imr-details/2380 [Accessed: 30 January 2022]

[117] Banihani SA. Tomato (*Solanum lycopersicum L.*) and type 2 diabetes. International Journal of Food Properties. 2018;**21**(1):99-105. DOI: 10.1080/10942912.2018.1439959

[118] Song Y, Cook NR, Albert CM, et al. Effects of vitamins C and E and beta-carotene on the risk of type 2 diabetes in women at high risk of cardiovascular disease: A randomized controlled trial. The American Journal of Clinical Nutrition. 2009;**90**(2):429-437. DOI: 10.3945/ajcn.2009.27491

[119] Kataja-Tuomola M, Sundell JR, Mannisto S, et al. Effect of alpha-tocopherol and beta-carotene supplementation on the incidence of type 2 diabetes. Diabetologia. 2008;**51**:47-53. DOI: 10.1007/s00125-007-0864-0

[120] Gao Q, Zhong C, Zhou X, et al. The association between intake of dietary lycopene and other carotenoids and gestational diabetes mellitus risk during mid-trimester: A cross-sectional study. British Journal of Nutrition. 2019;**121**(12):1405-1412. DOI: 10.1017/S0007114519000606

[121] Vinha AF, Barreira SVP, Costa ASG, et al. Pre-meal tomato (*Lycopersicon esculentum*) intake can have anti-obesity effects in young women? International Journal of Food Sciences and Nutrition. 2014;**65**(8):1019-1026. DOI: 10.3109/09637486.2014.950206

Chapter 4

Carotenoids in Cassava (*Manihot esculenta* Crantz)

Lovina I. Udoh, Josephine U. Agogbua, Eberechi R. Keyagha and Itorobong I. Nkanga

Abstract

Cassava is produced globally and consumed as an important staple in Africa for its calories, but the crop is deficient in micronutrients such as vitamin A. Pro-vitamin A carotenoids including β-carotene are precursors of vitamin A in the human body. Carotenoids are generally associated with colors of fruits and vegetables. Although most cassava varieties have white tuberous roots and generally accepted, naturally; some cassava roots are colored yellow and contain negligible amounts of vitamin A. Several genes have been identified in the carotenoids biosynthesis pathway of plants, but studies show that Phytoene synthase 2 (*PSY2*), lycopene epsilon cyclase, and β-carotene hydroxylase genes have higher expression levels in yellow cassava roots. So far, the *PSY2* gene has been identified as the key gene associated with carotenoids in cassava. Some initiatives are implementing conventional breeding to increase pro-vitamin A carotenoids in cassava roots, and much success has been achieved in this regard. This chapter highlights various prediction tools employed for carotenoid content in fresh cassava roots, including molecular marker-assisted strategies developed to fast-track the conventional breeding for increased carotenoids in cassava.

Keywords: cassava, carotenoids, marker assisted selection, molecular markers, vitamin A, biofortification, phenotyping

1. Introduction

Cassava (*Manihot esculenta* Crantz) is an important crop globally, and in Nigeria, it is consumed as a staple by more than 100 million people every day [1]. Global production of cassava has been given at approximately 278.7 million tons, it was estimated to be 281 million tons and 288.4 million tons in 2015 and 2016, respectively [2]. Global cassava market in 2019 increased by 0.4% to $164.1B, and consumption was peak at $172.1B [3]. Nigeria stands out as largest producer as its progressive cassava pattern increased from 42.5 million tons in 2010 to 61 million tons in 2020; total production area in 2012 was 3.85 million hectares [4]. In Nigeria, the cost of cassava production per hectare is estimated to be 82,055 naira, with a profit of about 123,745 naira. Although, in Africa, 50% of the cassava produced is largely consumed as food after processing; 38% in fresh and/or cooked form; and 12% is utilized for animal feed [5]. The crop is cultivated mostly by small scale farmers because it outperforms other staple food crops under long-term drought and poor soil conditions [6].

Commonly available white cassava can provide most of the body's daily energy needs, but it does not provide adequate protein, essential micronutrients, and vitamin A. Vitamin A deficiency makes the body susceptible to infection, especially among women and children [7]. It causes illness and eye defects that can lead to partial or complete blindness [7]. Most cultivars of cassava are white or off-white, and the roots of tubers are generally low in carotenoids [8]. Cassava varieties with colored pulp that may be rich in carotenoids are very rarely available and are not well known to the general public. Yellow flesh color of some cassava varieties is associated with the presence of carotenoids [9, 10], and the nutritive importance of carotenoids is attributed to its conversion to vitamin A when consumed. The consumption of tuberous roots of β-carotene-rich cultivars may contribute significantly to addressing vitamin A deficiency in sub-Saharan Africa.

One of the most important micronutrients with deficiency of high public health concern is vitamin A, followed by iron, zinc, and iodine [11]. The generic descriptor for compounds with the qualitative biological activity of retinol is vitamin A. It exists in the form of preformed retinoids that are preserved in animal tissues as pro-vitamin A carotenoids usually gotten from green, yellow, and/or orange plant tissues. A total of two-thirds of dietary vitamin A worldwide and more than 80% in the developing world have been said to come from carotenoids in vegetables [5]. The all-trans-β-carotene is observed to be the most abundant carotenoid in cassava together with isomers such as 9-cis β-carotene, 13 cis-β-carotene, and β-cryptoxanthin [5, 12, 13]. Several carotenoid biosynthesis genes and enzymes such as lycopene epsilon cyclase (*LCYε*), β-carotene hydroxylase (CHYβ), phytoene synthase 1 and 2, lycopene β-cyclase, and phytoene desaturase (*PDS*) have been identified for different plants including cassava [14–16]. Studies by Olayide et al. [13] detected more carotenoids and isomers in the leaves than roots. Phytoene synthase 2 (*PSY2*), *LCYε*, and CHYβ genes were mostly associated with β-carotene content in white and yellow roots, but they generally had higher expression in yellow root cassava [13]. To enhance marker-assisted selection in the conventional breeding to increase carotenoids in cassava roots, six single-nucleotide polymorphisms (SNP) markers were designed on candidate genes and validated on 650 elite cassava accessions of which PSY2_572 explained most of the phenotypic variation (R^2 = 0.75) in root pulp color [12].

Limited access to diets that are rich in vitamin A is known to be the root cause of vitamin A deficiency in Africa and other vitamin A deficiency inflicted regions. Efforts are continually being made to improve the nutritional value of cassava through biofortification, which has led to an improvement of its carotenoid content. These improvements have been successful through the adoption of advanced breeding techniques, which involves the screening of large numbers of genotypes for nutritional quality, agronomic traits, yield traits, etc., in order to select progenies with the best traits for further breeding.

2. Origin and domestication of cassava

Cassava (*Manihot esculenta* Crantz) is a dicotyledonous plant belonging to the Euphorbiaceous family known only in the cultivated form and was first domesticated by the Amerindians of South and Central America [17]. There is archaeological evidence of two major centers of origin for cassava, one in Mexico and Central America and the other in North-eastern Brazil. In sixteenth century, Portuguese navigators took cassava from Brazil to the western coast of Africa [18] and later to East Africa in eighteenth century through island of Reunion, Madagascar, also Zanzibar as described by Iglesias et al. [18]. It was introduced in India in the nineteenth century. Cassava plantations were set up by the Portuguese, who colonized South American regions by 1500 A.D. They carried cassava from these plantations to other continents [19]; hence, cassava was first introduced to Africa and Asia in the late sixteenth century by the Portuguese travelers. It was initially planted around the Congo River basin from where it moved to West and Central Africa [17, 20]. Nigeria was among the first African countries to receive the crop in the eighteenth century. The cassava crop was perhaps introduced in southern Nigeria by freed slaves who returned from South America through Sao Tome and Fernando Po islands [21].

2.1 Taxonomy

Cassava, as it is called in English, is referred to as "manioc" in French, "yuca" in Spanish, and "mandioca" in Portuguese. Cassava comprises about 7200 species. It belongs to the following [22];

Kingdom – Plantae
Subkingdom – Tracheobionta
Super division – Spermatophyta
Division – Magnoliophyta
Class – Magnoliopsida
Subclass – Rosidae
Order – Euphorbiales
Family – Euphorbiaceae
Subfamily – Manihotae
Genus – Manihot
Species – *Manihot esculenta* Crantz

This family is characterized by lactiferous vessels composed of secretory cells [17]. A total of 98 *Manihot* species have been recognized with one species (*Manihotoides pauciflora*) known in the closest related genus [17]. A lot of its characteristics have not been identified in any *Manihot* species, which are its mono-flower inflorescences and leaves borne at the apex of short, condensed stems arising from branch-lets. *M. pauciflora* is suggested to be a possible progenitor of all the Manihot groups. Unfortunately, this species is on the verge of extinction [23], and cassava is the only species that is widely cultivated for food production [23, 24]. The cultivated species may be derived from the wild progenitor *M. flabellifolia* [17].

2.2 Botanical description

Cassava is propagated mainly from stem cuttings, thereby maintaining true-to-type cultivars. Nevertheless, propagation by seed can take place naturally or during plant breeding procedures. When stem cuttings are planted in the moist soil under favorable conditions, they produce sprouts and adventitious roots at the base of the cuttings within a week. If propagated by seeds, it first develops into a tap root system. Cassava leaves are simple; it consists of a lamina and a petiole. Each leaf is subtended by two stipules, about 1 cm long. The petiole is between 5 and 30 cm long and varies from green to purple. The smooth margin of the lamina is palmate or lobed. The lobes differ in number, ranging from 3 to 9, and are most of the time odd numbers. The lobe's vain color can differ from green to purple. Most cassava varieties grown in Africa have elliptical or lanceolated lobes [17, 25]. The arrangement of cassava leaves on a stem (phyllotaxis) is a 2/5 spiral, meaning that the position of five leaves turns twice spirally around the stem, then the next leaf comes just above the beginning of the other. Their stems are cylindrical and have a diameter, which varies between 2 and 6 cm. Cassava stems usually grow up to 4 m, but some genotypes may grow to only to a height of1 m. The older parts of the stems display prominent knob-like scars, which are leaf scars and their nodes [20, 25]. Cassava is a monoecious plant with male and female flowers located on the same plant. The inflorescences are produced at the reproductive branches [22].

Cassava is propagated from stem cutting or seed. In cassava, the fleshy part is the central portion of the tuberous root. Tuberous roots vary in shape and color, depending on the soil conditions and variety [25]. Cassava grows between 30°N and 30°S in areas where annual rainfall is greater than 500 mm and where mean temperature is greater than 20 °C. However, some cassava varieties grow at 2000 m altitude or in subtropical areas with annual mean temperatures as low as 16 °C. Cassava prefers a sandy or sandy loam soil, but all types of soils, except water logged soils, can be used. Cassava tolerates the high levels of aluminum and manganese often found in tropical soils [26].

3. Carotenoids biosynthesis in plants

Exploitation of the diverse tropical cassava collection for development of high pro-vitamin A cassava cultivars entails understanding and application of knowledge derived from molecular and biochemical studies of carotenoids and their biosynthesis in plants. Carotenoids are naturally occurring organic pigments that are produced by plants and some photosynthetic organisms [27, 28]. They are characterized by their extensive conjugated double bond along their carbon backbone giving them the capability to absorb lights in the range of blue to green range of the visible spectrum [28]. In plants, carotenoids are present mainly as indispensable integral components of the chloroplast, providing multiple services to the photosynthetic machinery participating in the light harvesting process and guarding the photosystems from possible damages by quenching reactive singlet oxygens and radicals created during photooxidation [29–31].

The carotenoid biosynthesis pathway is extensively studied in plants [29–33] and is responsible for the biogenesis of about 600 40-carbon isoprenoid compounds broadly classified as xanthophylls and carotenes. The first reaction dedicated to siphoning substrates to the carotenoid biosynthesis pathway in plants is catalyzed by the enzyme phytoene synthase (*PSY*). In this reaction, two geranylgeranyl pyrophosphate molecules are condensed to produce the first colorless linear carotenoid compound, phytoene. Phytoene is then modified through a series of desaturation and isomerization reactions catalyzed by enzymes including phytoene desaturase (*PDS*) and carotenoid isomerase (*CRTISO*) yielding the red colored carotenoid, lycopene. Lycopene is the forking point in the pathway that leads to two separate downstream branches called α and β branches. In the α branch, carotenoids such as α-carotene and lutein are synthesized, while in the β branch, carotenoids such as β-carotene, β-cryptoxanthin, and zeaxanthin are generated following cyclization of the terminals of the linear structured lycopene. Key enzymes involved in the branched part of the pathway include lycopene epsilon α-cyclase (*LYCε*) and lycopene epsilon β cyclase

(LYCβ) and β-carotene hydroxylase. The *LYCβ* can add β-ionone rings in both ends of lycopene to give β-carotene; while *LYCε* can add ε-ring in one end only to give α-carotene [14, 30]. Hydroxylation at the C-3 position of each ring of β- carotene and α-carotene produces xanthophylls, zeaxanthin and lutein, respectively (**Figure 1**).

Among all carotenoid compounds, only β-carotene has full vitamin A activity due to its doubly ended β-ionone rings, while carotenoids that have single β ring, such as α-carotene and β-cryptoxanthin, have half vitamin A activity of β-carotene [18, 30, 34–36]. Although the mechanism of regulation of the carotenoid biosynthesis is still not fully understood, a lot of progress has been made in this regard [30, 37].

3.1 Genes associated with carotenoid in cassava

Studies by Iglesias and Chavez et al. [10, 18] reported that relatively few major genes are involved in the determination of carotenoid accumulation in cassava roots. Thus, the trait can be improved to a significant level through the process of selection and recombination. In other crops, genes such as phytoene synthase (PSY), β-carotene hydroxylase, lycopene β, and ε cyclase have been reported to play a role in increasing levels of carotenoids [36, 38]. In cassava, Arango et al. [39] observed three *PSY* genes, one of which was discovered to be associated with stress in the Poaceae homologs. However, the two remaining *PSY* genes contributed differentially to carotenoid accumulation in leaves, roots, and flower parts of cassava. So far, the *PSY* gene has been identified as the key gene associated with carotenoids in cassava [12, 40, 41]. Olayide et al. [13] observed that carotenoid synthesis genes were expressed in both white and yellow cassava roots, but the following genes had higher expression in yellow roots, including phytoene synthase 2, lycopene epsilon cyclase, and β-carotene hydroxylase.

3.2 Breeding for increased carotenoids in cassava roots

Cassava is an highly important diet not only for humans but also in animal diet especially poultry, due to its availability and calories [24, 42, 43]; thus, the need arose

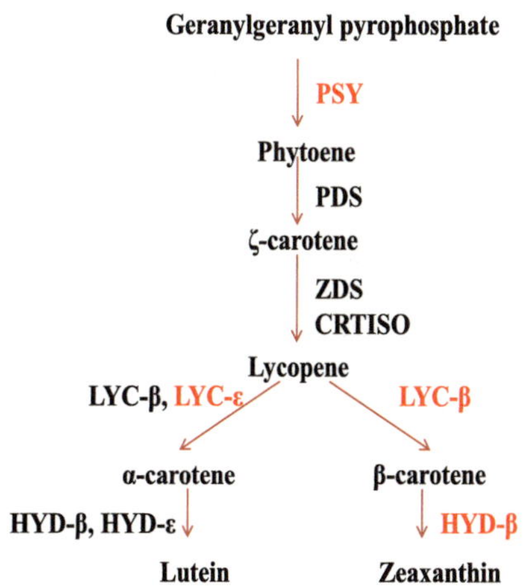

Figure 1.
A simplified diagram of the carotenoid biosynthetic pathway in plants, showing major genes and enzymes involved ***Figure 1****.*

to fortify the crop with micronutrient to improve its nutritional status. Some cassava varieties originally have yellow root color (**Figure 2**) meaning they have negligible amount of pro-vitamin A [18, 23]. Total carotenoid concentration in fresh yellow cassava ranges from 1 to 100 μg/g (fresh weight), primarily as all-trans-β-carotene, and is located in the parenchyma cells, the storage cells of the roots, and isomers such as 9 and 15 cis β-carotene and β-cryptoxanthin have also been detected [5, 18]. Carotenoid concentration is a stable trait and is influenced more by genotype than by its environment. Studies showed that retention of carotenoids differs not only per processing and storage method for a certain variety [10] but also within a variety, and this might be due to the variable distribution of dry weight matter within a root [44]. Retention varies between 10% for heavily processed and roasted cassava granules and 87% for boiling [10, 45].

Genetic improvement for this crop has employed crossing the wild yellow cultivars with elite breeding lines through recurrent selection and recombination [46]. This is accompanied by extensive field evaluation (phenotyping), including observations of disease and pest resistance, plant architecture, flowering ability, and performance in storage root [47]. Recently, rapid cycling recurrent selection was employed, which is able to cut down on the number of breeding cycles [9, 44]. The color of fruits and vegetables is associated with the presence of carotenoids, and the tuber-flesh color of some cassava accessions is yellow [23]. This indicates that naturally in the gene pool there are accessions with negligible amount of carotenoids [48], and this is currently being utilized in breeding. Breeding to biofortify cassava with pro-vitamin A will have a significant positive impact on nutrition and overall health, especially among poorer communities.

3.3 Comparison of carotenoids detected in some root crops

Using the high-performance liquid chromatography (HPLC), carotenoids and isomers have been detected in cassava, yam, and cocoyam, including the all trans-β carotene, 9-cis β carotene, 15-cis β carotene, β-cryptoxanthin (**Table 1**). From **Table 1**, all trans β-carotene was generally higher across cassava, with some accessions having up to 21 μg/g [5]. Also, the maximum value for total carotenoids content (TCC) in cassava as quantified using spectrophotometer is within the same range (12.95–14.8 μg/g). Yam

Figure 2.
Genetic variability of cassava cultivars with respect to carotenoid biosynthesis in storage roots. (A) Deep yellow roots, (B) white roots, (C) cream-colored root, (D) cassava plant, (E) unpeeled cassava roots ***Figure 2.***

Plant	Trait (μg/g)	Tool	n	Min	Max	SD	Mean	Reference
	βcryp	HPLC	252	0.01	1.93	0.15	0.1	[12]
			4074	0	3.77	0.12	0.14	[5]
Cassava	9cisβc	HPLC	252	0.02	2.77	0.49	0.72	[12]
M. esculenta			4895	0	3.95	0.54	0.99	[5]
	13cisβc	HPLC	252	0.02	2.06	0.31	0.5	[12]
			4920	0	2.24	0.6	1.01	[5]
	Alltransβc	HPLC	252	0.03	6.7	1.17	1.51	[12]
			4952	0	21	3.94	7.28	[5]
	TCC		252	0.07	10.14	1.79	2.73	[12]
			4952	0.11	29	5.08	11.51	[5]
	TCC	Spec	252	0.07	13.34	2.45	3.75	[12]
			35	2.87	12.95		7.99	[9]
			98	0.02	14.8			[15]
Yam	9cisβc	HPLC	1				1.93	[49]
D. cayenensis	13cisβc		1				0.34	[49]
	Alltransβc		1				2.83	[49]
	TCC		1				11.99	[49]
	9cisβc		1				0	[49]
D. dumetorum	13cisβc		1				0	[49]
	Alltransβc		1				0.59	[49]
	TCC		1				3.79	[49]
	9cisβc		1				1.14	[49]
D. bulbifera	13cisβc		1				0.02	[49]
	Alltransβc		1				0.27	[49]
	TCC		1				10.11	[49]
Cocoyam	9cisβc	HPLC	1				1.13	[49]
X. mafa (Scoth)	13cisβc		1				0.91	[49]
	Alltransβc		1				3.88	[49]
	TCC		1				14.87	[49]

Table 1.
Carotenoids from cassava and other tuber crops (fresh weight).

and cocoyam also had higher all trans β-carotene content across the accessions studied. The mean value of this carotenoid was higher in cassava, as stated in Ceballos et al. [5], compared with yam and cocoyam. Mean for TCC quantified by HPLC ranged from 2.73 to 11.51 μg/g. In yam, the TCC varied among the studied genotypes. The variety of cocoyam studied had high TCC (14.79 μg/g) compared with yam (11.99 μg/g).

4. Various prediction tools for carotenoids in Cassava

Selection for a trait can be made based on phenotypes or genotype using molecular tools. The physical outlook of organisms, which includes all seen and quantitative characters that can be accessed from the outer part of the plant, is the phenotype. This comprises attributes that provide structural phenotypic information such as counts, dimensions, colors, etc., as well as physiological attributes such as photosynthetic efficiencies, water content, surface properties, etc., resulting from genotype and environmental interactions [50].

Carotenoid phenotyping in cassava is very essential as it measures and quantifies its total carotene content. To ensure optimal quality of breeding programs, there must be an understanding of crop genotype interaction with the environment, and this is expressed by the proceeding genotypes and monitored by phenotyping [51]. As breeding for higher carotenoid levels in cassava advances, selection is a major drawback, as some means of predicting total carotenoid content may be really expensive such as the use of high-performance liquid chromatography.

Further, color intensity in cassava roots has been observed to be closely related to quantity of carotenoids in the roots [10]. While visual selection is useful for separating white from yellow root cassava, it cannot efficiently distinguish the salient differences between yellow roots. Other methods exist to quantify carotenoids or check color intensity such as the use of near-infrared spectroscopy (NIRS) [8], but here, we compare some frequently used phenotyping methods for carotenoids in cassava.

Different instruments employed to predict carotenoids in cassava roots are as itemized below:

4.1 Near-infrared spectroscopy (NIRS)

This technique measures the interrelationship between electromagnetic radiation and the vibrational properties of chemical bonds, which results in the absorption of part of the radiation energy. The visible spectra cover between 380 nm and 780 nm and capture mainly information on pigmentation due to the carotenoids present in the root [52]. NIRS aims to analyze a sample such as to get from it qualitative and quantitative information about its physical and chemical composition. This it does by treating spectra mathematically so as to obtain the relevant information in the spectra, which is connected to the character of interest [44]. Its principle of action involves calibration of the spectrometer in order to develop mathematical models that will connect the standard values to a linear combination of the values of absorbance. NIRS allows the timely screening of many samples and variables and measures samples in different states, i.e., both in solid and liquid forms. When compared with other phenotyping methods, it is a fast and nondestructive alternative for analyzing several constituents simultaneously while requiring minimal to no sample preparation. It is economical and possesses no hazard to the environment [53].

The NIRS provides quality phenotyping method for field-based breeding programs especially where there are no standard laboratories, therefore reducing the need to transport samples from the field while also cutting out the need for sample procession [8, 53]. In NIRS, calibration and data obtained can be shared between spectrometers, thus increasing the chances of developing a network of high-throughput phenotyping technique for screening cassava roots [9].

4.2 Chromameter

The chromameter is a tool for precise and objective assessment of surface color. It can be used to preselect materials for further analysis. It records data output in the form of the L* a* b* color coordinate. This system has been used for different studies pertaining to skin color [8, 54]. The L * corresponds to levels of darkness or lightness between black and white colors. Coordinate a* signifies the balance between red/ green, and b* between yellow/blue. This simple technique has equally been used to accurately quantify color intensity and quality in some plant tissues [13]. Sanchez et al. [8] observed that total carotenoid content and color intensity were strongly and positively associated ($R2 = 0.769$, $P < 0.01$), suggesting that the roots of cassava clones with a relatively high total carotenoid content can be selected through a simple visual inspection of the color intensity in the parenchyma. The difference in color of 228 biofortified cassava clones was also analyzed by [55], using the L* a* b* color coordinate system resulting in a high positive correlation between total carotenoids content (TCC) and the variables b* ($r = 0.90$) and chroma ($r = 0.89$). Their results demonstrate that the use of data obtained from this device is an economical, fast, and effective alternative for the development of TCC phenotyping tools with high predictive ability.

4.3 Image-based phenotyping

Digital image analysis allows the extraction of information regarding root color based on the strong correlation that exists between digital and virtual data [55]. Imaging techniques possess high resolutions, which permit the visualization of the sample from several dimensions and generating multiple data. Image-based phenotyping is used to quantify complex plant characters such as growth pattern, photosynthetic abilities, yield, tolerance to biotic and abiotic stress, both in controlled environments and in the open field. Plants imaging aims to measure a character quantitatively through the interaction that takes place between light and the plant such as reflection, absorption, and transmission of sent photons of which all plant cells and tissue possess specific wavelength for light reflection, absorption, and transmission. Since the presence of carotenoid is linked with the intensity of yellow color, it is taken that this type of phenotyping is ideal for the quantification of root carotenoid content. There are different aspects to image-based phenotyping, and they include thermal infrared imaging, imaging spectroscopy, fluorescence imaging, visible imaging, laser imaging, and hyperspectral imaging [55]. The advantages of imaging techniques include the following:

I. It is time saving.

II. Commercially available digital cameras that are easy to handle, transport and open-source software for processing images can be used.

III. It gives room for thorough reexamination of images recorded in cases where doubts arise concerning the phenotyping process.

IV. Calibration of prediction models makes it possible for sample size to be reduced, thus concentrating on samples of greatest interest, thereby reducing cost.

4.4 iCheck Carotene

This is a portable device consisting of two components, namely the measuring unit (iCheck™ Carotene) and the disposable reagent vial (iEx™) where the reaction is performed. The disposable reagent vial contains 2 mL of a mixture of reagents, which is needed for carrying out the reaction. The iCheck Carotene is very portable weighing about 250 g with dimensions (200 mm x 104 mm x 40 mm) making it easily transportable. It uses rechargeable batteries, which can be used to take up to about 400 measurements, which saves automatically and can be retrieved at will as a text file with the use of a USB cable. The iCheck Carotene is a rapid screening method, which is cost-effective, user-friendly, simple, and inexpensive. It does not require highly skilled and specialized personnel for its operation, neither does it need an expensive laboratory setup with equipment and specified chemicals; therefore, it is suitable for the quantification of a large number of samples within a short period of time with accurate results especially where there are no labs available, and there is a large number of cassava genotypes to be screened [56].

4.5 High-performance liquid chromatography (HPLC)

This is an advanced form of liquid chromatography, which is used in the separation, identification, and quantification of components in a mixture of molecules encountered in chemical and biological systems. It is associated with high reproducibility, ease of selection, manipulation, and high rate of recovery [57]. Its working principle involves a solution of the sample being injected into a column of a porous material (stationary phase) while a liquid (mobile phase) is pumped at high pressure into the column. The sample separates based on the differences in the rates of migration through the column, which results from the partitioning of the sample between the stationary and the mobile phase [57, 58].

In cassava phenotyping, HPLC is used in the separation and quantification of individual carotenoids, which are different in their provitamin A activity. Although it has high reproducibility, its analysis is expensive, costing 50–70 US dollars per sample with very low throughput. It is time-consuming, labor-intensive, and requires a highly sophisticated laboratory setup with highly skilled personnel and strictly adhered quality control regiment [57].

4.6 Ultraviolet–visible (UV–vis) spectrophotometer

The UV–Visible Spectrophotometer is a type of spectrophotometer, principle of which is based on the absorption of ultraviolet light or visible light by chemical compounds, and this results in the production of distinct spectra. It is a device that precisely measures electromagnetic energy at specific wavelengths of lights. UV–visible spectrophotometer uses light over the ultraviolet range of (185–400 nm) and visible range (400–700 nm) of the electromagnetic radiation spectrum. Carotenoids concentration, for example, is determined spectrophotometrically by measuring the absorbance (also referred to as optical density) of the extract at various wavelengths. The absorption spectrum of β-carotene (carotenoids) peaks between 450 and 475 nm. UV spectrophotometer has been mostly used to quantify carotenoids in cassava and other plants. Jaramillo et al. observed that spectrophotometer reading gave a higher quantity of total carotenoids content (30.0 μg/g) compared with the use of iCheck devise (24.7 μg/g). Other authors have also quantified carotenoids in cassava using the spectrophotometer [5, 12, 15, 57]. The major throwback with the use of this instrument is that it is cumbersome and time-consuming with low throughput especially when dealing with large breeding populations.

5. Marker-assisted selection of carotenoid-rich cassava

Over the years, conventional breeding has been augmented by various innovative molecular marker-aided techniques. Genetic differences that exist between individual species and organisms represent a genetic marker. Generally, they do not represent the target genes themselves but act as "signposts" or "landmarks" representing DNA along chromosomes. The first marker technologies involved the use of biochemical markers such as isozymes and allozymes. These gave way to the first-generation DNA markers such as restriction fragment length polymorphism (RFLP), random amplified polymorphic DNA (RAPD), amplified fragment length polymorphism (AFLP), and simple sequence repeat (SSR). Advances in sequencing technology enhanced the use of DNA-sequencing based markers such as SSR and SNP, giving rise to automated high-throughput genotyping [59]. For a genetic marker to be useful, the marker locus has to show experimentally detectable variation among individuals [15, 60]. The variation can be due to single-nucleotide polymorphisms or deletions/insertions or major chromosomal changes. Molecular genetic markers can be used to study the diversity of the observable variation at population or species level [59]. They can also be used to map genomes, identify regions of the genome controlling a trait, and follow a segment of interest of the genome in a plant breeding scheme [59–61].

Molecular markers are usually utilized in a breeding program to facilitate and speed up the selection process as such, carotenoids are boosted through marker-assisted selection (MAS) on target genes [62]. Some of the applications of molecular markers such as RFLP, AFLP, RAPD, SSR in cassava include taxonomical studies, understanding the phylogenetic relationships in the genus, confirmation of ploidy, genetic diversity assessment, and genetic mapping studies in cassava [59], making MAS a reality for application in breeding programs [63]. SSRs have also been used to select for carotenoids in cassava [64]. The reduced cost of the new technologies increases the discovery and utilization of new set of molecular markers that is amenable for the high-throughput genotyping [65].

Recently, single-nucleotide polymorphism (SNP) markers are increasingly being used for genotyping to study gene function. SNPs work as molecular markers that help locate genes associated with a trait and are used for genotype sequencing. SNPs may play a direct role in a trait and affect gene function if they occur within a gene or in a regulatory coding region and thus serve as molecular markers. These markers can be applied in the following: genetic architecture detection, association studies, conservation genetics, genetic diversity, and are fast becoming the marker system of choice in marker assisted plant breeding programs. Some genotyping methods that can specifically genotype an SNP affecting a trait in a collection of population include the use of KASP (competitive allele-specific polymerase chain reaction (PCR) markers, especially for a small number of SNPs [65, 66]. It utilizes a unique form of competitive allele-specific PCR combined with a novel, homogeneous, fluorescence-based reporting system for the identification and measurement of genetic variation occurring at the nucleotide level to detect single-nucleotide polymorphisms (SNPs) or inserts and deletions (InDels) [36, 38]. KASP chemistry provides a versatile choice that can be applied to small- and large-scale projects. It is suitable for use on a variety of equipment platforms and provides flexibility in terms of the number of SNPs and

Marker	MAF	Het	PIC	Trait	*p* value	Marker R^2
PSY2_572	0.81	0.16	0.26	Color chart	3.26×10^{-198}	0.75
				b^*	7.38×10^{-199}	0.78
				TC SPEC	1.95×10^{-20}	0.62
				TBC	3.96×10^{-18}	0.57
PSY2_549	0.76	0.25	0.3	Color chart	3.64×10^{-146}	0.63
				b^*	7.77×10^{-120}	0.59
				TC SPEC	1.08×10^{-19}	0.59
				TBC	5.58×10^{-17}	0.54
lcyE_1066	0.73	0.32	0.32	TBC	9.63×10^{-04}	0.13
				TC SPEC	0.00322	0.11
				Color chart	0.00262	0.02
				b^*	0.01152	0.01
lcyE_1294	0.98	0.04	0.04	Color chart	3.93×10^{-06}	0.03
				b^*	1.94×10^{-07}	0.04
lcyE_1015	0.96	0.05	0.07	Color chart	1.36×10^{-04}	0.03
				b^*	8.86×10^{-06}	0.04
lcyE_829	0.82	0.19	0.21	Color chart	0.01377	0.01

**MAF–major allele frequency, Het–heterozygosity, PIC–polymorphic information content, Chromameter b*, PSY2–Phytoene synthase2 gene, lcyE–Lycopene epsilon cylase gene, Pulpcol–pulp-color score, TC SPEC–total carotenoid by spectrophotometer, TC–iCheck total carotenoid by iCheck Fluoro, TBC–Total β-carotene.*
*Source: **Table 2** [12].*

Table 2.
Summary results of validated SNP markers on cassava breeding collection.

the number of samples able to be analyzed. To facilitate the selection of carotenoid-rich cassava genotypes, six KASP SNP markers were designed on candidate genes and validated on 650 elite cassava accessions of which PSY2_572 explained most of the phenotypic variation (R^2 = 0.75) in root pulp color (**Table 2**) [12].

Most recent advances in next-generation sequencing technologies have enabled the use of genome-wide SNP markers for genomic selection. The genomic selection tool is believed to significantly increase the efficiency of breeding by increasing the speed and accuracy of selection in a breeding program by predicting the genetic value of individuals at an early selection stage [67]. Genomic selection models have also been implemented by [68], to fast-track the improvement of provitamin A carotenoids in cassava using a total of 23,431 single-nucleotide polymorphic markers.

6. Conclusion

Cassava (*Manihot esculenta* Crantz) is produced globally and a food security crop for many households in sub-Saharan Africa. Commonly available white cassava can provide most of the body's daily energy needs, but it is deficient in vitamin A and some essential micronutrients such as iron and zinc. Vitamin A deficiency makes the body susceptible to infection. Most widely consumed cultivars of cassava are white or off-white, and the roots are generally low in carotenoids [10]. Although recently, some yellow pulp-colored varieties associated with the presence of carotenoids are being propagated, and it is gradually gaining public acceptance. The nutritive importance of carotenoids is attributed to its conversion to vitamin A when consumed. Numerous genes have been identified in the carotenoids biosynthesis pathway of plants, but studies show that phytoene synthase 2 (*PSY2*), lycopene epsilon cyclase, and β-carotene hydroxylase genes have higher expression levels in yellow cassava roots. So far, the *PSY2* gene has been identified as the key gene associated with increased carotenoids in cassava and has also been tested for its efficiency in breeding [12, 41].

One bottleneck associated with the breeding for increased carotenoids in cassava storage roots is the phenotyping as large populations need to be subjected to selection. The most highly reproducible tool in predicting carotenoids is the high-performance liquid chromatography (HPLC), but its analysis is expensive, costing 50–70 US dollars per sample with very low throughput [57]. Thus, other easy-to-use devices have been accessed for use in phenotyping carotenoids in cassava such as the near-infrared spectroscopy, Chromameter, iCheck Carotene device. These devices have been observed to have high correlation with HPLC, for instance, total β-carotene as quantified by HPLC had high correlation ($r = 0.75$) with total carotenoids quantified using the iCheck device [12].

Also, molecular markers tools such as simple sequence repeats, single-nucleotide polymorphisms and even genomic selection [12, 64, 68] have been employed to speed up the breeding for increased carotenoids in cassava roots.

Acknowledgements

We acknowledge the support of Mr. A. I. Udoh.

Conflict of interest

No conflict of interest.

Notes/thanks/other declarations

Thanks to all coauthors for contributing to the success of this work.

Appendices and Nomenclature

NIRS	Near-infrared spectroscopy
HPLC	High-performance liquid chromatography
TCC	Total carotenoid content
PSY2	Phytoene synthase 2
LCYE	Lycopene epsilon cyclase
MAS	Marker-assisted selection
SSRs	Simple sequence repeats

Author details

Lovina I. Udoh[1]*, Josephine U. Agogbua[2], Eberechi R. Keyagha[3] and Itorobong I. Nkanga[1]

1 Faculty of Biological Science, Department of Botany, Akwa Ibom State University, Ikot-Akpaden, Akwa Ibom State, Nigeria

2 Department of Plant Science and Biotechnology, Faculty of Science, University of Port Harcourt, Rivers State, Nigeria

3 Department of Crop Science and Technology, School of Agriculture and Agricultural Technology, Federal University of Technology, Owerri, Imo State, Nigeria

*Address all correspondence to: lovinaudoh@aksu.edu.ng

References

[1] Onyemauwa CS. Analysis of Household Consumption of Cassava Products in Ohaozara, Ebonyi State, Southeast Nigeria. Research. 2010;**2**:1-6

[2] FAO: Agriculture: Why cassava. FAO. 2009. Available from: http://www.fao.org/ag/agp/agpc/gcds/index_en.html. [Accessed 24 September 2016]

[3] Global Cassava Market Is Expected to Successfully Resist the COVID Pandemic. Global Trade Magazine. Available from: https://www.globaltrademag.com/global-cassava-market-is-expected-to-successfully-resist-the-covid-pandemic/. [Accessed 17 April 2022]

[4] World Food and Agriculture – Statistical Yearbook 2021. World Food Agric – Stat Yearb 2021. Epub ahead of print 28 October 2021. DOI: 10.4060/CB4477EN

[5] Ceballos H, Davrieux F, Talsma EF, et al. Carotenoids in Cassava Roots. Carotenoids. Epub ahead of print 14 June 2017. DOI: 10.5772/INTECHOPEN.68279.

[6] Esuma W, Herselman L, Labuschagne MT, et al. Genome-wide association mapping of provitamin A carotenoid content in cassava. Euphytica. Epub ahead of print 9 September 2016. DOI: 10.1007/s10681-016-1772-5

[7] Saltzman A, Birol E, Bouis HE, et al. Biofortification: Progress toward a more nourishing future. Global Food Security. 2013;**2**:9-17

[8] Sánchez T, Ceballos H, Dufour D, et al. Prediction of carotenoids, cyanide and dry matter contents in fresh cassava root using NIRS and Hunter color techniques. Food Chemistry. 2014;**151**:444-451

[9] Ortiz D, Sánchez T, Morante N, et al. Sampling strategies for proper quantification of carotenoid content in cassava breeding. Journal of Plant Breeding and Crop Science. 2011;**3**:14-23

[10] Chávez AL, Sánchez T, Jaramillo G, et al. Variation of quality traits in cassava roots evaluated in landraces and improved clones. Euphytica. 2005;**143**:125-133

[11] Owens BF, Gore MA, Magallanes-Lundback M, et al. A foundation for provitamin a biofortification of maize: Genome-wide association and genomic prediction models of carotenoid levels. Genetics. 2014;**198**:1699-1716

[12] Udoh LI, Gedil M, Parkes EY, et al. Candidate gene sequencing and validation of SNP markers linked to carotenoid content in cassava (Manihot esculenta Crantz). Mol Breed; 37. Epub ahead of print 2017. DOI: 10.1007/s11032-017-0718-5

[13] Olayide P, Large A, Stridh L, et al. Gene Expression and Metabolite Profiling of Thirteen Nigerian Cassava Landraces to Elucidate Starch and Carotenoid Composition. Agronomy. 2020;**10**:424

[14] Carvalho LJ, Agustini MA, Anderson JV, et al. Natural variation in expression of genes associated with carotenoid biosynthesis and accumulation in cassava (Manihot esculenta Crantz) storage root. BMC Plant Biology. 2016;**16**:133

[15] Luo X, Tomlins KI, Carvalho LJCB, et al. The analysis of candidate genes and loci involved with carotenoid metabolism in cassava (Manihot esculenta Crantz)

using SLAF-seq. Acta Physiologiae Plantarum. 2018;**40**:1-11

[16] Nisar N, Li L, Lu S, et al. Carotenoid Metabolism in Plants. Molecular Plant. 2015;**8**:68-82

[17] Isendahl C. The Domestication and Early Spread of Manioc (Manihot Esculenta Crantz): A Brief Synthesis. Latin American Antiquity. 2011;**22**:452-468

[18] Iglesias C, Mayer J, Chavez L, et al. Genetic potential and stability of carotene content in cassava roots. Euphytica. 1997;**94**:367-373

[19] Manu-Aduening JA, Lamboll RI, Ampong Mensah G, et al. Development of superior cassava cultivars in Ghana by farmers and scientists: The process adopted, outcomes and contributions and changed roles of different stakeholders. Euphytica. 2006;**150**:47-61

[20] Olsen KM, Schaal BA. Evidence on the origin of cassava: Phylogeography of Manihot esculenta. Proceedings of the National Academy of Sciences of the United States of America. 1999;**96**:5586

[21] The spread of cassava (manioc) in Igboland, south-east Nigeria: a reappraisal of the evidence on JSTOR, Available from: https://www.jstor.org/stable/23317131. [Accessed 29 March 2022]

[22] Rival L, McKey D. Domestication and diversity in manioc (Manihot esculenta Crantz ssp. esculenta, Euphorbiaceae). Current Anthropology. 2008;**49**:1119-1128

[23] Nassar NMA, Junior OP, Sousa MV, et al. Improving carotenoids and amino-acids in cassava. Recent Patents on Food, Nutrition & Agriculture. 2009;**1**:32-38

[24] Cock JH. Cassava: A Basic Energy Source in the Tropics. Science (80-). 1982;**218**:755-762

[25] Alves AAC. Cassava botany and physiology. Cassava: Biology, Production and Utilization. 2002:67-89

[26] Costa AG, Da L, Souza S, et al. Characterization of soils cultivated with cassava under different managements. Brazilian Journal of Agricultural and Environmental Engineering Caracterização de solos cultivados com mandioca sob diferentes manejos. 2021:764-771

[27] Beyene G, Solomon FR, Chauhan RD, et al. Provitamin A biofortification of cassava enhances shelf life but reduces dry matter content of storage roots due to altered carbon partitioning into starch. Plant Biotechnology Journal;**16**. Epub ahead of print 2018. DOI: 10.1111/pbi.12862

[28] Cazzonelli CI. Goldacre Review: Carotenoids in nature: insights from plants and beyond. Functional Plant Biology. 2011;**38**:833

[29] Tracewell CA, Vrettos JS, Bautista JA, et al. Carotenoid Photooxidation in Photosystem II. Archives of Biochemistry and Biophysics. 2001;**385**:61-69

[30] Sun T, Rao S, Zhou X, et al. Plant carotenoids: recent advances and future perspectives. Molecular Horticulture. 2022;**2**:1-21

[31] Bang H, Yi G, Kim S, et al. Watermelon lycopene β-cyclase: promoter characterization leads to the development of a PCR marker for allelic selection. Euphytica. 2014;**200**:363-378

[32] Giuliano G. Plant carotenoids: genomics meets multi-gene engineering.

Current Opinion in Plant Biology. 2014;**19**:111-117

[33] Udoh LI, Gedil M, Parkes EY. Candidate gene sequencing and validation of SNP markers linked to carotenoid content in cassava (Manihot esculenta Crantz). Epub ahead of print 2017. DOI: 10.1007/s11032-017-0718-5

[34] Giuliano G. Provitamin A biofortification of crop plants: a gold rush with many miners. Current Opinion in Biotechnology. 2017;**44**:169-180

[35] Li F, Murillo C, Wurtzel ET. Maize Y9 Encodes a Product Essential for 15-cis-ζ-Carotene Isomerization. Plant Physiology. 2007;**144**:1181-1189

[36] Harjes CE, Rocheford TR, Bai L, et al. Natural genetic variation in lycopene epsilon cyclase tapped for maize biofortification. Science. 2008;**319**:330-333

[37] Cunningham FX, Gantt E. Genes and enzymes of carotenoid biosynthesis in plants. Annual Review of Plant Physiology and Plant Molecular Biology. 1998;**49**:557-583

[38] Yan J, Kandianis CB, Harjes CE, et al. Rare genetic variation at Zea mays crtRB1 increases β-carotene in maize grain. Nature Genetics. 2010;**42**:322-327

[39] Arango J, Wüst F, Beyer P, et al. Characterization of phytoene synthases from cassava and their involvement in abiotic stress-mediated responses. Planta. 2010;**232**:1251-1262

[40] Welsch R, Arango J, Bär C, et al. Provitamin A accumulation in cassava (Manihot esculenta) roots driven by a single nucleotide polymorphism in a phytoene synthase gene. Plant Cell. 2010;**22**:3348-3356

[41] Ige A, Olasanmi B, Bauchet GJ, et al. Validation of SNP markers for marker-assisted selection of genotypes with increased carotenoid and dry matter contents in cassava. bioRxiv Prepr. 2022:1-23

[42] Morgan NK, Choct M. Cassava: Nutrient composition and nutritive value in poultry diets. Animal Nutrition. 2016;**2**:253-261

[43] Tonukari NJ. Cassava and the future of starch. Electronic Journal of Biotechnology 2004; 7: 0-0.

[44] Ceballos H, Morante N, Sanchez T, et al. Rapid cycling recurrent selection for increased carotenoids content in cassava roots. Crop Science. 2013;**53**:2342-2351

[45] Taleon V, Sumbu D, Muzhingi T, et al. Carotenoids retention in biofortified yellow cassava processed with traditional African methods. Journal of the Science of Food and Agriculture. 2019;**99**:1434-1441

[46] Ssemakula G, Dixon AGO, Maziya-Dixon B. Stability of total carotenoid concentration and fresh yield of selected yellow-fleshed cassava (Manihot esculenta Crantz). Journal of Tropical Agriculture. 2007;**45**:14-20

[47] Okechukwu RU, AGO D. Genetic Gains from 30 Years of Cassava Breeding in Nigeria for Storage Root Yield and Disease Resistance in Elite Cassava Genotypes. Journal of Crop Improvement. 2008;**22**:181-208

[48] Bradbury JH et al. Chemistry of Tropical Root Crops: Significance for Nutrition and Agriculture in the Pacific. Australian Centre for International Agricultural Research Canberra; 1988

[49] Ukom AN, Ojimelukwe PC, Ezeama CF, et al. Proximate composition

and carotenoid profile of yams (Dioscorea spp.) and cocoyam (Xanthosoma maffa (Scoth)) root tubers from Nigeria. Department of Food Science and Technology, Michael Okpara University of Agriculture; 2014. pp. 1-10

[50] Mackay TFC. Question & Answer Q & A: Genetic analysis of quantitative traits. Journal of Biology:8

[51] Alamu EO, Nuwamanya E, Cornet D, et al. Near-infrared spectroscopy applications for high-throughput phenotyping for cassava and yam: A review. International Journal of Food Science and Technology. 2021;**56**:1491

[52] Pellicer A, del C Bravo M. Near-infrared spectroscopy: A methodology-focused review. Seminars in Fetal & Neonatal Medicine. 2011;**16**:42-49

[53] Abincha W, Ikeogu UN, Kawuki R, et al. Portable spectroscopy calibration with inexpensive and simple sampling reference alternatives for dry matter and total carotenoid contents in cassava roots. Applied Sciences. 2021;**11**:1-11

[54] Abincha W, Kawuki R, Ikeogu U, et al. Comparison of Near-infrared Spectroscopy with other options for total carotenoids content phenotyping in fresh cassava roots. Journal of the Science of Food and Agriculture. 2020;**4**:55-60

[55] de Carvalho RRB, Cortes DFM, e Sousa MB, et al. Image-based phenotyping of cassava roots for diversity studies and carotenoids prediction. PLoS One. 2022;**17**:e0263326

[56] Maroya NG, Kulakow PA, Parkes EY, et al. Quantification of total carotene content of yellow root cassava genotypes using iCheck TM Carotene in comparison with the standard spectrophotometer method Contact. 2006;**57881**

[57] Jaramillo AM, Londoño LF, Orozco JC, et al. A comparison study of five different methods to measure carotenoids in biofortified yellow cassava (Manihot esculenta). PLoS One:13. Epub ahead of print 1 December 2018. DOI: 10.1371/journal.pone.0209702

[58] Belalcazar J, Dufour D, Andersson MS, et al. High-throughput phenotyping and improvements in breeding cassava for increased carotenoids in the roots. Crop Science. 2016;**56**:2916-2925

[59] Ferguson M, Rabbi I, Kim D-J, et al. Molecular Markers and Their Application to Cassava Breeding: Past, Present and Future. Tropical Plant Biology. 2012;**5**:95-109

[60] Okogbenin E, Porto MCM, Egesi C, et al. Marker-Assisted Introgression of Resistance to Cassava Mosaic Disease into Latin American Germplasm for the Genetic Improvement of Cassava in Africa. Crop Science. 2007;**47**:1895

[61] Bernardo AN, Bowden RL, Rouse MN, et al. Validation of molecular markers for new stem rust resistance genes in U.S. hard winter wheat. Crop Science. 2013;**53**:755-764

[62] Tanksley SD, Orton TJ. Isozyme in Plant Genetics and Breeding.

[63] Chen X, Sullivan PF. Single nucleotide polymorphism genotyping: biochemistry, protocol, cost and throughput. The Pharmacogenomics Journal. 2003;**3**:77-96

[64] Njoku DN, Gracen VE, Offei SK, et al. Parent-offspring regression analysis for total carotenoids and some agronomic traits in cassava. Euphytica. 2015;**206**:657-666

[65] Yang S, Yu W, Wei X, et al. An extended KASP-SNP resource for

molecular breeding in Chinese cabbage (Brassica rapa L. ssp. pekinensis). PLoS One. 2020;**15**:1-17

[66] Zhao S, Li A, Li C, et al. Development and application of KASP marker for high throughput detection of AhFAD2 mutation in peanut. Electronic Journal of Biotechnology. 2017;**25**:9-12

[67] Crossa J, Pérez P, Hickey J, et al. Genomic prediction in CIMMYT maize and wheat breeding programs. Heredity (Edinb):112. Epub ahead of print 2014. DOI: 10.1038/hdy.2013.16

[68] Esuma W, Ozimati A, Kulakow P, et al. Effectiveness of genomic selection for improving provitamin A carotenoid content and associated traits in cassava. G3 Genes|Genomes|Genetics:11. Epub ahead of print 6 September 2021. DOI: 10.1093/G3JOURNAL/JKAB160

Section 3

Carotenoids Characterization

Chapter 5

Computational Chemistry Study of Natural Apocarotenoids and Their Synthetic Glycopeptide Conjugates as Therapeutic Drugs

Norma Flores-Holguín, Juan Frau and Daniel Glossman-Mitnik

Abstract

The objective of the research to be presented in the chapter is the determination of the chemical reactivity properties of some natural apocarotenoids and their synthetic glycopeptide conjugates that could have the ability to inhibit SARS-CoV-2 replication. The study will be based on the consideration of the Conceptual DFT branch of Density Functional Theory (DFT) through the consideration of particular successful model chemistry which has been demonstrated as satisfying the Janak and Ionization Energy theorems within Generalized Gradient Approximation (GGA) theory. The research will be complemented by a report of the ADMET and pharmacokinetic properties hoping that this information could be of help in the development of new pharmaceutical drugs for fighting COVID-19.

Keywords: natural Apocarotenoids, glycopeptide conjugates, computational chemistry, SARS-CoV-2, COVID-19, chemical reactivity, conceptual DFT

1. Introduction

Cyclic peptides have several desirable qualities, including high binding affinity, target selectivity, and low toxicity, which make them a promising therapeutic development approach. Antimicrobial peptides (AMPs), also known as host defense peptides, are short, positively charged peptides found in a wide range of life forms, including microbes and humans. The majority of AMPs are capable of directly killing microbial infections, whereas some operate indirectly by altering the host defensive mechanisms [1].

Teicoplanin, a therapeutically utilized glycopeptide antibiotic, has surfaced as a possible antiviral in the context of the global COVID-19 pandemic, with its potency being increased with lipophilic modifications. Teicoplanin was obtained from *Actinoplanes teichomyceticus*, which was recovered from a soil sample collected in Nimodi Village, Indore, India, for the first time in 1978. Teicoplanin's structure was discovered in 1984. Teicoplanin has been identified as a lipoglycopeptide antibiotic. This antibiotic is made up of a heptapeptide made up of seven aromatic amino acids, sugar residues, and a lipid chain that is nonribosomal. It is made up of five identical chemicals that differ in their fatty acid side chains and are generated by bacteria [2]. This glycopeptide antibiotic typically used to treat Gram-positive bacterial infections has been demonstrated to diminish SARS-CoV-1 and MERS-CoV infection [3].

Lipophilic modifications have been shown to improve the antiviral spectrum and efficacy of glycopeptide antibiotics, which improve antiviral activity against coronaviruses, HIV, flavivirus, and influenza viruses with the drawback of being associated with substantial cytotoxicity [4–13]. To obtain efficient glycopeptide antibiotics by increasing their lipophilicity and avoiding the cytotoxicity problems, recent research has been presented with a study of the structural and biochemical properties of new lipophilic apocarotenoid conjugates of Teicoplanin and its pseudoaglycone [14].

Inspired by this latest research and as a follow up of our previous studies on the chemical reactivity properties of carotenoids [15–19] and cyclopeptides [20–24], we think that it is worth reporting the physicochemical and bioactivity properties of some of these apocarotenoid conjugates of Teicoplanin as well as to predict and understand their chemical reactivity properties considering a methodology developed by our research group. This will be done as a means of further validation of the procedure and for assessing the behavior of the MN12SX density functional in the fulfillment of the Janak theorem and the Ionization Energy Theorem, which is a corollary of the former [25–29].

Thus, the objective of this work is to report the results of a computational study of the bioactivity properties and chemical reactivity of three apocarotenoid conjugates of Teicoplanin based on the CDFT-based Computational Peptidology (CDFT-CP) methodology [20–24]. These three molecules will be designed by considering the Teicoplanin A3–1 variant (PubChem CID 15122170) and the apocarotenoids Bixin, Methylcrocetin and -apo-8'-Carotenoic Acid. The methodology will be based on the combination of the chemical reactivity descriptors from Conceptual Density Functional Theory (CDFT) [30–35] with some Cheminformatics tools [36–43] which may be utilized to assess the associated physicochemical properties. This will be complemented by the detection of the ability of the three molecules to act as possible useful drugs through an analysis of their bioactivities and pharmacokinetics characteristics linked to the ADMET features [44–46].

2. Methodology

2.1 Density functional theory calculations

The Kohn-Sham (KS) methodology approach to Density Functional Theory (DFT) involves the determination of the electronic density, the molecular energy, and the orbital energies of a specific system, in particular, the HOMO and LUMO frontier orbitals which are intrinsically related to the chemical reactivity of the molecules [47–50]. The definitions for the global reactivity descriptors that form the core of Conceptual DFT are [30–35]:

$$\text{Electronegativity} \quad \chi \approx \frac{1}{2}(\varepsilon_L + \varepsilon_H) \tag{1}$$

$$\text{Global Hardness} \quad \eta \approx (\varepsilon_L - \varepsilon_H) \tag{2}$$

$$\text{Electrophilicity} \quad \omega \approx (\varepsilon_L + \varepsilon_H)^2 / 4(\varepsilon_L - \varepsilon_H) \tag{3}$$

$$\text{Electrodonating Power} \quad \omega^- \approx (3\varepsilon_H + \varepsilon_L)^2 / 16\eta \tag{4}$$

$$\text{Electroaccepting Power} \quad \omega^+ \approx (\varepsilon_H + 3\varepsilon_L)^2 / 16\eta \tag{5}$$

$$\text{Net Electrophilicity} \quad \Delta\omega^{\pm} = \omega^+ + \omega^- \tag{6}$$

being ε_H and ε_L the frontier orbital energies related to the molecular systems considered in this research. These global reactivity descriptors that arise from Conceptual DFT [30–35], have been complemented by the estimation of the Nucleophilicity Index N [51–55] that takes into account the value of the HOMO energy obtained using the KS scheme using an arbitrary shift of the origin with tetracyanoethylene (TCE) as a reference.

Conformational analysis of the studied molecules has been achieved using MarvinView 17.15 from ChemAxon [http://www.chemaxon.com], which was applied to undertake Molecular Mechanics calculations considering the MMFF94 force field [56–60]. This was followed in each case by a geometry optimization and frequency calculation using the Density Functional Tight Binding (DFTB) methodology [61]. This last step was required for the verification of the absence of imaginary frequencies as a confirmation of the stability of every optimized structure as being a minimum in the energy surface. The determination of the electronic properties and the Conceptual DFT reactivity descriptors of the studied molecules was addressed through the MN12SX/Def2TZVP/H2O model chemistry [62–64] because it has been previously shown that it verifies the KID procedure fulfiling the Ionization Energy Theorem, with the help of the Gaussian 16 software [61] and the context of the SMD solvation model [65]. The charge of all the molecules was taken as equal to zero whereas the radical anion and cation were considered in the doublet spin state. The SMD solvation model was chosen because it has been shown that it provides atomic charges of the Hirshfeld kind that are almost independent of the basis set and which are usually recommended for calculations within Conceptual Density Functional Theory.

2.2 Computational pharmacokinetics and ADMET report

The SMILES notation of each studied molecule was generated through the Online SMILES Translator and Structure File Generator [https://cactus.nci.nih.gov/translate/], and then was fed into the online program Chemicalize from ChemAxon [http://www.chemaxon.com], which was considered to get a glimpse of the potential therapeutic properties of the studied molecular systems (accessed: January 2022).

A similarity search in the chemical space of compounds with molecular structures that could be compared to the ones being studied, with already known biological and pharmacological properties, was achieved through the online Molinspiration software from Molinspiration Cheminformatics [https://www.molinspiration.com/] (accessed, January 2022).

Pharmacokinetics is a procedure that involves determining the likely fate of a medicinal molecule in the body, which is critical information in the creation of new medicine. Individual indices named Absorption, Distribution, Metabolism, Excretion, and Toxicity (ADMET) factors have typically been used to analyze the associated consequences. Chemicalize and the internet available pkCSM, a software for the prediction of small-molecule pharmacokinetic properties using SMILES, was also used to obtain additional information regarding the Pharmacokinetics parameters and ADMET indices [45].

3. Results and discussion

3.1 Conceptual DFT-based computational peptidology

The optimized molecular structures of the three apocarotenoid glycopeptide conjugates considered through this research through the methodology presented before are displayed in **Figure 1**:

The quality of the chosen density functional may be realized by comparing its results with results from high-level computations or experiential values. Nevertheless, this comparison is not always computationally practicable because of the large size of the molecules or the lack of experimental results for the chemical methods being explored. Our research group has developed a methodology known as KID [20–24], as an aid to evaluating a particular density functional about its internal coherence. It is evident that within the Generalized Kohn-Sham (GKS) version of DFT, some relationships exist between the KID methodology and the Ionization Energy Theorem, which is a corollary of Janak theorem [25–29]. This is done by connecting ε_H to -I and ε_L to -A, through

$$J_I = \varepsilon_H + E_{gs}(N-1) - E_{gs}(N) \quad (7)$$

$$J_A = \varepsilon_L + E_{gs}(N) - E_{gs}(N+1) \quad (8)$$

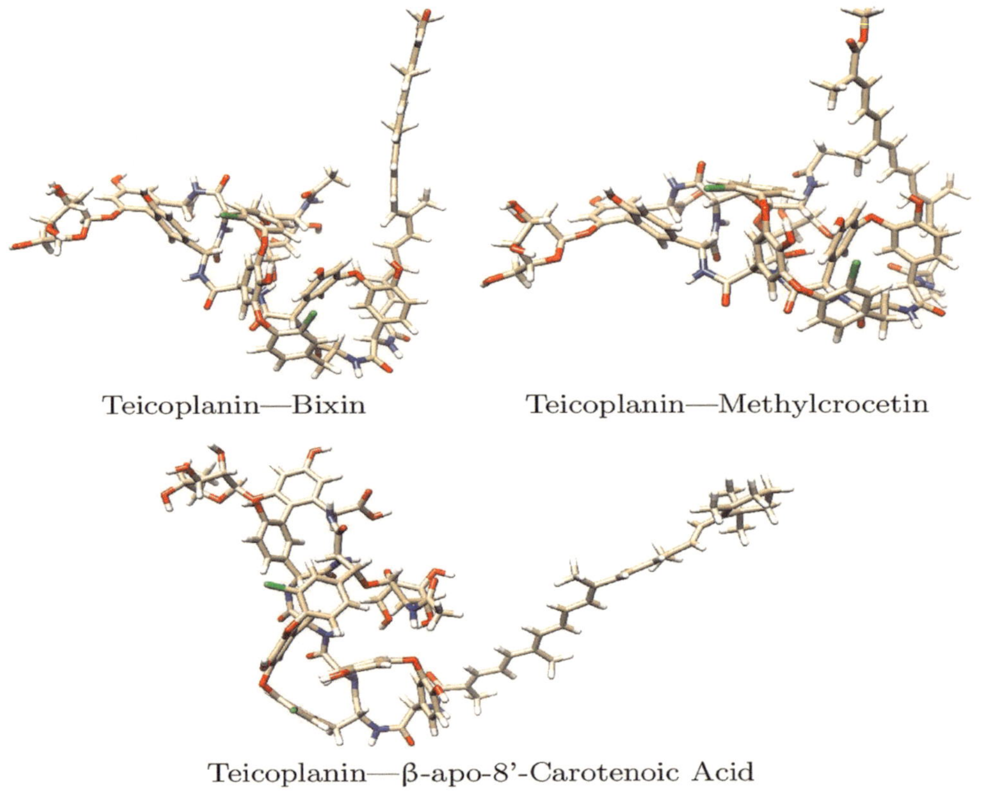

Figure 1.
Optimized molecular structures of three apocarotenoid glycopeptide conjugates (Brown: C, blue: N, red: O, green: Cl, and white: H).

$$J_{HL} = \sqrt{{J_I}^2 + {J_A}^2} \tag{9}$$

Another KID descriptor ΔSL related to the difference in energies between the SOMO and the LUMO of the neutral system has been devised to aid in the verification of the accuracy of the methodology.

The MN12SX density functional has been shown to have a Koopmans-compliant behavior in earlier studies of the chemical reactivity of diverse molecular systems. However, for further validation of this model chemistry in the prediction of the chemical reactivity properties of the apocarotenoid glycopeptides conjugates considered here, additional research is necessary. The CDFT software tool was used to make this determination, and the findings are shown in **Table 1**:

The results from **Table 1** are very interesting because they show that there is an almost perfect fulfillment of the Janak and Ionization Energy theorems for the MN12SX/Def2TZVP/H2O model chemistry employed in this work.

Having verified that the MN12SX/Def2TZVP/H2O is the most adequate one for obtaining accurate results for the Conceptual DFT global reactivity descriptors, the estimated values for the Global Reactivity Descriptors (including the Nucleophilicity N) for the three molecular systems acquired utilizing the mentioned CDFT tool are displayed in **Table 2**:

The electronegativity (χ) and global hardness (η) are absolute values for the chemical reactivity that have not a known experimental counterpart. Indeed, they can be estimated by resorting to the experimental vertical ionization energy (I) and vertical electron affinity (A) but these values are not known for the molecule under study. A different thing can be said about electrophilicity ω and Nucleophilicity (N). The electrophilicity ω index involves a compromise between the tendency of an electrophile to acquire extra electron density and its resistance to exchange electron

Molecule	HOMO	LUMO	SOMO	H-L Gap	J(I)	J(A)	J(HL)	ΔSL
1	−5.246	−2.693	−2.701	2.553	0.001	0.004	0.005	0.008
2	−5.262	−2.711	−2.717	2.552	0.003	0.005	0.005	0.007
3	−5.073	−2.377	−2.389	2.696	0.002	0.007	0.008	0.012

1: Teicoplanin-Bixin; 2: Teicoplanin-Methylcrocetin; 3: Teicoplanin-β-apo-8′-Carotenoic Acid.

Table 1.
Frontier orbital energies, H-L gap and the KID indices (all in eV) were used for the verification of the ionization energy theorem behavior of the MN12SX density functional in the study of the chemical reactivity of the synthetic conjugates of the glycopeptide Teicoplanin with several apocarotenoids.

Molecule	χ	η	ω	S	N	ω^-	ω^+	$\Delta\omega\pm$
1	3.970	2.553	3.086	0.392	3.546	8.316	4.346	12.662
2	3.987	2.552	3.114	0.392	3.530	8.380	4.394	12.774
3	3.725	2.696	2.574	0.371	3.720	7.178	3.453	10.632

1: Teicoplanin-Bixin; 2: Teicoplanin-Methylcrocetin; 3: Teicoplanin-β-apo-8′-Carotenoic Acid.

Table 2.
Global reactivity descriptors for the synthetic conjugates of the glycopeptide Teicoplanin with several apocarotenoids: Electronegativity (χ), hardness (η), Electrophilicity (ω) (all in eV), softness S (in eV^{-1}), Nucleophilicity N, Electrodonating power (ω^-), Electroaccepting power (ω^+) and net Electrophilicity ($\Delta\omega\pm$) (also in eV).

density with the environment [55]. By considering a group of Diels-Alder reactions and the electrophiles involved in them [53, 66, 67], classification of organic compounds as strong, moderate, or marginal electrophiles, that is an electrophilicity ω scale, was established, with ω larger than 1.5 eV for the first instance, with ω between 0.8 and 1.5 eV for the second case, and ω smaller than 0.8 eV for the final case [53, 66, 67]. By checking **Table 2**, it can be said that the three molecules may be regarded as strong electrophiles. Domingo and his collaborators [51–55] have also proposed a Nucleophilicity index N through the consideration of the HOMO energy obtained through the KS scheme with an arbitrary shift of the origin taking the molecule of tetracyanoethylene (TCE) as a reference. An analysis of a series of common nucleophilic species participating in polar organic reactions allowed them to establish a further classification of organic molecules as strong nucleophiles with $N > 3.0$ eV, moderate nucleophiles with $2.0 < N < 3.0$ eV and marginal nucleophiles with
$N < 2.0$ eV. By checking again **Table 2**, it can be concluded that the three molecular systems may be considered also as strong nucleophiles.

It is interesting to see that in comparison with similar research with peptides [20–24], the MN12SX/Def2TZVP/H2O model chemistry retains its predictive ability even when the glycopeptides are conjugated with carotenoids, as in the present case. An important point is that the conjugates are predicted to be strong nucleophiles and electrophiles while the computed behavior for isolated peptides depicts them as moderate or even marginal nucleophiles and electrophiles.

3.2 Computational pharmacokinetics and ADMET report

The majority of medicinal drugs work by attaching to target protein molecules while at the same time modifying their functions. The Bioactivity Scores, which are a measure of the capacity of the molecules to act or coordinate with distinct receptors, are listed in **Table 3** for the three apocarotenoid glycopeptide conjugates:

These bioactivity scores for organic molecules can be interpreted as active (when the bioactivity score is greater thanÂ 0), moderately active (when the bioactivity score lies between −5.0 and 0.0) and inactive (when the bioactivity score is lower than −5.0).

The pharmacokinetics of a drug is evaluated through ADMET research, which is acronymous for Absorption, Distribution, Metabolism, Excretion, and Toxicity. If absorption is unsatisfactory, the distribution and metabolism of the drug would be changed, potentially resulting in nephrotoxicity and neurotoxicity. As a result, ADMET analysis is one of the most important aspects of computational drug design. In addition to the previous Conceptual DFT-based Computational Peptidology and Pharmacokinetics results, we are complementing this study with a report of the computed ADMET features as shown in **Table 4**:

Molecule	GPCR ligand	Ion channel modulator	Nuclear receptor ligand	Kinase inhibitor	Protease inhibitor	Enzyme inhibitor
1	−4.08	−4.11	−4.12	−4.11	−4.07	−4.08
2	−4.08	−4.10	−4.11	−4.10	−4.06	−4.07
3	−4.08	−4.09	−4.13	−4.11	−4.08	−4.08

1: Teicoplanin-Bixin; 2: Teicoplanin-Methylcrocetin; 3: Teicoplanin-β-apo-8'-Carotenoic Acid.

Table 3.
Bioactivity scores of the synthetic conjugates of the glycopeptide Teicoplanin with several apocarotenoids.

Property	MOL 1	MOL 2	MOL 3
		Absorption	
Water Solubility (log mol/L)	−2.892	−2.892	−2.892
Caco2 Permeability (log Papp 10^{-6} cm/s)	−0.965	−0.915	−0.748
Gastrointestinal Absorption (human) (% Absorbed)	8.189	6.874	25.006
Skin Permeability (log Kp)	−2.735	−2.735	−2.735
P-glycoprotein Substrate	Yes	Yes	Yes
P-glycoprotein I Inhibitor	No	No	No
P-glycoprotein II Inhibitor	No	No	No
		Distribution	
VDss (human) (log L/kg)	0.052	0.042	0.045
Fraction Unbound (human) (Fu)	0.363	0.368	0.367
BBB Permeability (log BB)	−5.180	−5.192	−4.950
CNS Permeability (log PS)	−7.097	−7.187	−6.499
		Metabolism	
CYP2D6 Substrate	No	No	No
CYP3A4 Substrate	No	No	No
CYP1A2 Inhibitor	No	No	No
CYP2C19 Inhibitor	No	No	No
CYP2C9 Inhibitor	No	No	No
CYP2D6 Inhibitor	No	No	No
CYP3A4 Inhibitor	No	No	No
		Excretion	
Total Clearance (log ml/min/kg)	−0.989	−1.037	−1.319
Renal OCT2 Substrate	No	No	No
		Toxicity	
AMES Toxicity	No	No	No
Max. Tolerated Dose (human) (log mg/kg/day)	0.438	0.438	0.438
hERG I inhibitor	No	No	No
hERG II inhibitor	No	No	No
Oral Rat Acute Toxicity (LD50) /mol/kg)	2.482	2.482	2.482
Oral Rat Chronic Toxicity (LOAEL) (log mg/kg-bw/day)	17.306	17.065	17.008
Hepatotoxicity	No	No	No
Skin Sensitization	No	No	No
T. Pyriformis Toxicity (log g/L)	0.285	0.285	0.285
Minnow Toxicity (log mM)	25.665	26.019	24.477

1: Teicoplanin-Bixin; 2: Teicoplanin-Methylcrocetin; 3: Teicoplanin-β-apo-8'-Carotenoic Acid.

Table 4.
Computed ADMET features of the synthetic conjugates of the glycopeptide Teicoplanin with several apocarotenoids.

It is important to note that all the members of the group of studied molecules display positive values for the Human Gastrointestinal Absorption (HI), in particular for MOL3, and negative values for the AMES toxicity and Hepatotoxicity. All the molecular systems will be P-glycoprotein inhibitors (P-gp), being also P-gp substrates. None of the apocarotenoid glycopeptide conjugates will be inhibitors of the molecules related to cytochrome P450, displaying also a negative behavior as substrates of the CYP2D6 and CYP3A4 variants. Finally, all the molecular systems considered here will display a negative result regarding their behavior as hERG inhibitors. These results are comparatively similar to those presented within the study of the structural and biochemical properties of lipophilic apocarotenoid conjugates of Teicoplanin and its pseudoaglycone that inspired this research [14].

4. Conclusions

The chemical reactivities of three apocarotenoid glycopeptide conjugates have been thoroughly investigated by optimizing their structures using the DFTB methodology and calculating their electronic properties using high-quality model chemistry, namely MN12SX/Def2TZVP/H2O. This model chemistry was already used in previous research, demonstrating its utility for this type of calculation. However, an involved estimation of the KID descriptors for all the molecules demonstrated the ability of the MN12SX density functional for the accurate estimation of the frontier orbital energies based on the KID procedure methodology. The fact that the energy of the LUMO and the SOMO (or the HOMO energy of the anion) are almost the same, which is reflected in the KID accuracy descriptor ΔSL being very close to zero, is an indication that the derivative discontinuity is negligible for the chosen density functional. This is translated as the ability of the LUMO energy to reflect with precision the Electron Affinity of the molecule, implying that the chemical reactivity parameters obtained by considering this density functional will be very accurate. This is a very important result because it allowed the estimation of the accuracy of the results based only on the fulfillment of some intrinsic requirements (like the Janak and Ionization Energies) without the need to resort to the comparison with experimental results that could not be available, as in the present case.

By considering our suggested Conceptual DFT-based Computational Peptidology methodology, the three apocarotenoid glycopeptide conjugates have been studied by applying certain techniques generally used in the procedure of drug discovery and development, showing that these molecular systems may be regarded as potential therapeutic drugs. The biological targets, physicochemical attributes, and ADMET (Absorption, Distribution, Metabolism, Excretion, and Toxicity) indices associated with their bioavailability and pharmacokinetics were forecasted and analyzed as descriptors that could be useful in future drug development research.

It may be concluded that the results coming from the present study may be of importance for the pharmaceutic industry because they show that the proposed three apocarotenoid glycopeptide conjugates fulfilled the objective of increasing the lipophilicity while at the same time avoiding the risk of the associated toxicity.

Acknowledgements

NFH and DGM are researchers of CIMAV and CONACYT (Mexico) and want to thank both institutions for partial support.

Conflict of interest

The authors declare no conflict of interest regarding the publication of this manuscript.

Author details

Norma Flores-Holguín[1†], Juan Frau[2†] and Daniel Glossman-Mitnik[1*†]

1 Laboratorio Virtual NANOCOSMOS, Departamento de Medio Ambiente y Energía, Centro de Investigación en Materiales Avanzados, Chihuahua, Mexico

2 Departament de Química, Universitat de les Illes Balears, Palma de Mallorca, Spain

*Address all correspondence to: daniel.glossman@cimav.edu.mx

† These authors contributed equally.

References

[1] Mahlapuu M, Håkansson J, Ringstad L, Björn C. Antimicrobial peptides: An emerging category of therapeutic agents. Frontiers in Cellular and Infection Microbiology. 2016;**6**:194

[2] Vimberg V. Teicoplanin – A new use for an old drug in the COVID-19 era? Pharmaceuticals. 2021;**14**(12):1227

[3] Schütz D, Ruiz-Blanco YB, Münch J, Kirchhoff F, Sanchez-Garcia E, Müller JA. Peptide and peptide-based inhibitors of SARS-CoV-2 entry. Advanced Drug Delivery Reviews. 2020;**167**:47-65

[4] Balzarini J, Keyaerts E, Vijgen L, Egberink H, De Clercq E, Van Ranst M, et al. Inhibition of feline (FIPV) and human (SARS) coronavirus by semisynthetic derivatives of Glycopeptide antibiotics. Antiviral Research. 2006;**72**(1):20-33

[5] Szűcs Z, Naesens L, Stevaert A, Ostorházi E, Batta G, Herczegh P, et al. Reprogramming of the antibacterial drug vancomycin results in potent antiviral agents devoid of antibacterial activity. Pharmaceuticals. 2020;**13**(7):139

[6] Bereczki I, Csávás M, Szűcs Z, Rőth E, Batta G, Ostorházi E, et al. Synthesis of antiviral perfluoroalkyl derivatives of teicoplanin and vancomycin. ChemMedChem. 2020; **15**(17):1661-1671

[7] Balzarini J, Pannecouque C, De Clercq E, Pavlov AY, Printsevskaya SS, Miroshnikova OV, et al. Antiretroviral activity of semisynthetic derivatives of glycopeptide antibiotics. Journal of Medicinal Chemistry. 2003;**46**(13): 2755-2764

[8] De Burghgraeve T, Kaptein SJF, Ayala-Nunez NV, Mondotte JA, Pastorino B, Printsevskaya SS, et al. An analogue of the antibiotic teicoplanin prevents flavivirus entry in vitro. PLoS One. 2012;**7**(5):e37244

[9] Naesens L, Vanderlinden E, Rőth E, Jekö J, Andrei G, Snoeck R, et al. Anti-influenza virus activity and structure-activity relationship of aglycoristocetin derivatives with cyclobutenedione carrying hydrophobic chains. Antiviral Research. 2009;**82**(1):89-94

[10] Pintér G, Batta G, Kéki S, Mándi A, Komáromi I, Takács-Novák K, et al. Diazo transfer-click reaction route to new, lipophilic teicoplanin and ristocetin aglycon derivatives with high antibacterial and anti-influenza virus activity: An aggregation and receptor binding study. Journal of Medicinal Chemistry. 2009;**52**(19):6053-6061

[11] Sipos A, Máté G, Rőth E, Borbás A, Batta G, Bereczki I, et al. Synthesis of fluorescent ristocetin aglycon derivatives with remarkable antibacterial and antiviral activities. European Journal of Medicinal Chemistry. 2012;**58**:361-367

[12] Bereczki I, Kicsák M, Dobray L, Borbás A, Batta G, Kéki S, et al. Semisynthetic teicoplanin derivatives as new influenza virus binding inhibitors: Synthesis and antiviral studies. Bioorganic & Medicinal Chemistry Letters. 2014;**24**(15):3251-3254

[13] Szűcs Z, Kelemen V, Le Thai S, Csávás M, Rőth E, Batta G, et al. Structure-activity relationship studies of lipophilic teicoplanin pseudoaglycon derivatives as new anti-influenza virus agents. European Journal of Medicinal Chemistry. 2018;**157**:1017-1030

[14] Bereczki I, Papp H, Kuczmog A, Madai M, Nagy V, Agócs A, et al. Natural

apocarotenoids and their synthetic glycopeptide conjugates inhibit SARS-CoV-2 replication. Pharmaceuticals. 2021;**14**(11):1111

[15] Flores-Holguín N, Frau J, Glossman-Mitnik D. Chemical reactivity properties, solubilities, and bioactivity scores of some pigments derived from carotenoids of marine origin through conceptual DFT descriptors. Journal of Chemistry. 2019;**1–12**:2019

[16] Flores-Hidalgo M, Torres-Rivas F, Monzón-Bensojo J, Escobedo-Bretado M, Glossman-Mitnik D, Barraza-Jiménez D. Electronic structure of carotenoids in natural and artificial photosynthesis. In: Cvetkovic DJ, Nikolic GS, editors. Carotenoids. Rijeka: InTech; 2017. pp. 17-33

[17] Ruiz-Anchondo T, Glossman-Mitnik D. Computational molecular characterization of the β,β-carotene molecule. Journal of Molecular Structure: THEOCHEM. 2009;**913**:215-220

[18] Ruiz-Anchondo T, Flores-Holguín N, Glossman-Mitnik D. Natural carotenoids as precursors of nanomaterials for molecular photovoltaics: A computational DFT study. Molecules. 2010;**15**(7): 4490-4510

[19] Payán-Gómez SA, Flores-Holguín N, Pérez-Hernández A, Pinón-Miramontes M, Glossman-Mitnik D. Computational molecular characterization of the flavonoid morin and its Pt(II), Pd(II) and Zn(II) complexes. Journal of Molecular Modeling. 2011;**17**(5):979-985

[20] Flores-Holguín N, Frau J, Glossman-Mitnik D. A fast and simple evaluation of the chemical reactivity properties of the Pristinamycin family of antimicrobial peptides. Chemical Physics Letters. 2020;**739**:137021

[21] Flores-Holguín N, Frau J, Glossman-Mitnik D. Conceptual DFT-based computational peptidology of marine natural compounds: Discodermins A–H. Molecules. 2020;**25**(18):4158

[22] Flores-Holguín N, Frau J, Glossman-Mitnik D. Virtual screening of marine natural compounds by means of chemoinformatics and CDFT-based computational peptidology. Marine Drugs. 2020;**18**(9):478

[23] Norma Flores-Holguín, Juan Frau, and Daniel Glossman-Mitnik. Conceptual DFT as a helpful chemoinformatics tool for the study of the Clavanin family of antimicrobial marine peptides. In Sergio Ricardo De Lazaro, Luis Henrique Da Silveira Lacerda, and Renan Augusto Pontes Ribeiro, editors, Density Functional Theory, Chapter 3. London,UK: Intech Open; 2021. pp. 57-67

[24] Flores-Holguín N, Frau J, Glossman-Mitnik D. A CDFT-based computational peptidology (CDFT-CP) study of the chemical reactivity and bioactivity of the marine-derived alternaramide cyclopentadepsipeptide. Journal of Chemistry. 2021;**1–11**:2021

[25] Janak JF. Proof that $\partial E/\partial n_i = \varepsilon$ in density functional theory. Physical Review B. 1978;**18**:7165-7168

[26] Kar R, Song J-W, Hirao K. Long-range corrected functionals satisfy Koopmans' theorem: Calculation of correlation and relaxation energies. Journal of Computational Chemistry. 2013;**34**(11):958-964

[27] Tsuneda T, Song J-W, Suzuki S, Hirao K. On Koopmans' theorem in density functional theory. The Journal of Chemical Physics. 2010;**133**(17):174101

[28] Tsuneda T, Hirao K. Long-range correction for density functional theory.

Wiley Interdisciplinary Reviews: Computational Molecular Science. 2014; **4**(4):375-390

[29] Kanchanakungwankul S, Truhlar DG. Examination of how well long-range-corrected density functionals satisfy the ionization energy theorem. Journal of Chemical Theory and Computation. 2021;**17**(8):4823-4830

[30] Parr RG, Yang W. Density-Functional Theory of Atoms and Molecules. New York: Oxford University Press; 1989

[31] Chermette H. Chemical reactivity indexes in density functional theory. Journal of Computational Chemistry. 1999;**20**:129-154

[32] Geerlings P, De Proft F, Langenaeker W. Conceptual density functional theory. Chemical Reviews. 2003;**103**:1793-1873

[33] Gázquez JL, Cedillo A, Vela A. Electrodonating and electroaccepting powers. Journal of Physical Chemistry A. 2007;**111**(10):1966-1970

[34] Chattaraj PK, Chakraborty A, Giri S. Net electrophilicity. Journal of Physical Chemistry A. 2009;**113**(37):10068-10074

[35] Geerlings P, Chamorro E, Chattaraj PK, De Proft F, Gázquez JL, Liu S, et al. Conceptual density functional theory: Status, prospects, issues. Theoretical Chemistry Accounts. 2020;**139**(2):36

[36] Engel T, Gasteiger J, editors. Applied Chemoinformatics : Achievements and Future Opportunities. Weinheim, Germany: Wiley-VCH; 2018

[37] Engel T, Gasteiger J, editors. Chemoinformatics: Basic Concepts and Methods. Weinheim: Wiley-VCH; 2018

[38] Bajorath J, editor. Chemoinformatics for Drug Discove. Hoboken, New Jersey: WILEY, A John Wiley & Sons Publication; 2014

[39] Varnek A, Tropsha A, editors. Chemoinformatics Approaches to Virtual Screening. Cambridge, UK: Royal Society of Chemistry; 2008

[40] Guha R, Bender A, editors. Computational Approaches in Cheminformatics and Bioinformatics. Hoboken, N.J: Wiley; 2012

[41] Benjamin B. Basic Principles of Drug Discovery and Development. Amsterdam Netherlands: Academic Press; 2015

[42] Medina-Franco JL, Saldívar-González FI. Cheminformatics to characterize pharmacologically active natural products. Biomolecules. 2020; **10**(11):11

[43] Firdaus Begam B, Satheesh Kumar J. A study on cheminformatics and its applications on modern drug discovery. Procedia Engineering. 2012;**38**:1264-1275

[44] Daina A, Michielin O, Zoete V. SwissADME: A free web tool to evaluate pharmacokinetics, drug-likeness and medicinal chemistry friendliness of small molecules. Scientific Reports. 2017;**7**(1): 42717

[45] Pires DEV, Blundell TL, Ascher DB. pkCSM: Predicting small-molecule pharmacokinetic and toxicity properties using graph-based signatures. Journal of Medicinal Chemistry. 2015;**58**(9): 4066-4072

[46] Chakraborty A, Pan S, Chattaraj PK. Biological activity and toxicity: A conceptual DFT approach. In: Structure and Bonding. Berlin Heidelberg: Springer; 2012. pp. 143-179

[47] Lewars E. Computational Chemistry - Introduction to the Theory and Applications of Molecular and Quantum Mechanics. Dordrecht: Kluwer Academic Publishers; 2003

[48] Young DC. Computational Chemistry - A Practical Guide for Applying Techniques to Real-World Problems. New York: John Wiley & Sons; 2001

[49] Jensen F. Introduction to Computational Chemistry. 2nd ed. Chichester, England: John Wiley & Sons; 2007

[50] Cramer CJ. Essentials of Computational Chemistry - Theories and Models. 2nd ed. Chichester, England: John Wiley & Sons; 2004

[51] Domingo LR, Chamorro E, Perez P. Understanding the reactivity of captodative ethylenes in polar cycloaddition reactions. A theoretical study. The Journal of Organic Chemistry. 2008;**73**(12):4615-4624

[52] Jaramillo P, Domingo LR, Chamorro E, Pérez P. A further exploration of a nucleophilicity index based on the gas-phase ionization potentials. Journal of Molecular Structure: Theochem. 2008;**865**(1–3): 68-72

[53] Domingo LR, Sáez JA. Understanding the mechanism of polar Diels-Alder reactions. Organic and Biomolecular Chemistry. 2009;7(17): 3576-3583

[54] Domingo LR, Perez P. The nucleophilicity N index in organic chemistry. Organic and Biomolecular Chemistry. 2011;**9**:7168-7175

[55] Domingo LR, Ríos-Gutiérrez M, Pérez P. Applications of the conceptual density functional theory indices to organic chemistry reactivity. Molecules. 2016;**21**:748

[56] Halgren TA. Merck molecular force field. I. Basis, form, scope, parameterization, and performance of MMFF94. Journal of Computational Chemistry. 1996;**17**(5–6):490-519

[57] Halgren TA. Merck molecular force field. II. MMFF94 van der Waals and electrostatic parameters for intermolecular interactions. Journal of Computational Chemistry. 1996;**17**(5–6): 520-552

[58] Halgren TA. MMFF VI. MMFF94s option for energy minimization studies. Journal of Computational Chemistry. 1999;**20**(7):720-729

[59] Halgren TA, Nachbar RB. Merck molecular force field. IV. Conformational energies and geometries for MMFF94. Journal of Computational Chemistry. 1996;**17**(5–6):587-615

[60] Halgren TA. Merck molecular force field. V. Extension of MMFF94 using experimental data, additional computational data, and empirical rules. Journal of Computational Chemistry. 1996;**17**(5–6):616-641

[61] Frisch MJ, Trucks GW, Schlegel HB, Scuseria GE, Robb MA, Cheeseman JR, et al. Gaussian 16 Revision C.01. Wallingford CT: Gaussian Inc; 2016

[62] Peverati R, Truhlar DG. Screened-exchange density functionals with broad accuracy for chemistry and solid-state physics. Physical Chemistry Chemical Physics. 2012;**14**(47):16187-16191

[63] Weigend F, Ahlrichs R. Balanced basis sets of Split valence, triple zeta valence and quadruple zeta valence quality for H to Rn: Design and

assessment of accuracy. Physical Chemistry Chemical Physics. 2005;**7**: 3297-3305

[64] Weigend F. Accurate coulomb-fitting basis sets for H to R. Physical Chemistry Chemical Physics. 2006;**8**: 1057-1065

[65] Marenich AV, Cramer CJ, Truhlar DG. Universal solvation model based on solute electron density and a continuum model of the solvent defined by the bulk dielectric constant and atomic surface tensions. Journal of Physical Chemistry B. 2009;**18**(1): 6378-6396

[66] Domingo LR, José Aurell M, Pérez P, Contreras R. Quantitative characterization of the global electrophilicity power of common diene/ dienophile pairs in Diels-Alder reactions. Tetrahedron. 2002;**58**(22):4417-4423

[67] Pérez P, Domingo LR, José Aurell M, Contreras R. Quantitative characterization of the global electrophilicity pattern of some reagents involved in 1,3-dipolar cycloaddition reactions. Tetrahedron. 2003;**59**(17): 3117-3125

Section 4

Carotenoids and Diseases

Chapter 6

Role of Carotenoids in Cardiovascular Disease

Arslan Ahmad, Sakhawat Riaz,
Muhammad Shahzaib Nadeem, Umber Mubeen
and Khadija Maham

Abstract

Carotenes are fat-soluble pigments found in a variety of foods, the majority of which are fruits and vegetables. They may have antioxidant biological properties due to their chemical makeup and relationship to cellular membranes. And over 700 carotenoids have been found, with—carotene, lutein, lycopene, and zeaxanthin is the most significant antioxidant food pigments. Their capacity to absorb lipid peroxides, reactive oxygen species (ROS) and nitrous oxide is likely linked to their anti-oxidative properties (NO). The daily requirements for carotenoids are also discussed in this chapter. Heart disease is still a prominent source of sickness and mortality in modern societies. Natural antioxidants contained in fruits and vegetables, such as lycopene, a-carotene, and B-carotene, may help prevent CVD by reducing oxidative stress, which is a major factor in the disease's progression. Numerous epidemiological studies have backed up the idea that antioxidants might be utilized to prevent and perhaps treat cardiovascular illnesses at a low cost. Supplements containing carotenoids are also available, and their effectiveness has been proven. This article provides an overview of carotenoids' chemistry, including uptake, transport, availability, metabolism, and antioxidant activity, including its involvement with disease prevention, notably cardiovascular disease.

Keywords: carotenoids, antioxidants, cardiovascular disease, free radicals, CVD prevention

1. Introduction

There are presently around 700 carotenoids known, although only about 50 of them are being digested by humans [1, 2]. Carotenoids are present in large concentrations in adipose tissue (80–80% of total), liver (8–12%), and muscles (2–3%) in healthy adults, but in fewer amounts in all other areas [3]. overall amount and levels of various carotenoids inside a person's bloodstream are mostly determined by their daily average diet. Carotenoids and polyenes are abundant in green leafy vegetables and various multicolored fruits [4]. The bulk of dietary carotenoids is digested by the stomach and enters the bloodstream in humans. People's blood contains B-carotene, a-carotene, cryptoxanthin, lycopene, and lutein [5]. Carotenoids circulate in the circulation alongside lipoproteins, notably LDL (low-density lipoprotein fraction) [6]. However, a large amount of ingested Β-carotene and other provitamins. A carotenoid is transformed to the retina, primarily in the gut wall, but also some proportion in the stomach and intestines [7]. In the human diet, fresh vegetables are currently the primary source of carotenoids [8–10]. Lutein might perform an important role in hypertension and symptoms of acute permeability in those with heart problems, high cholesterol, and/or hyperglycemia, according to a literature review and meta-analysis [11].

Carotenoids may be found in a variety of fruits and are also available as a nutritional supplement [12, 13]. Cardiovascular abnormalities have subsequently been a major source of worry across the world since they affect a large portion of the global population, and an elevated death rate has been reported in individuals aged 30 and above [14, 15]. Numerous researches have looked at carotenoids' possible cardioprotective and antioxidant capabilities [16, 17]. Individuals with cardiovascular disease may benefit from the anti-inflammatory properties of lutein, which may help to alleviate their symptoms [18]. ROS-induced reactive damage can arise in lipid peroxidation products, this may hasten the onset of atherosclerosis, the condition that causes heart attacks and ischemic strokes [19].

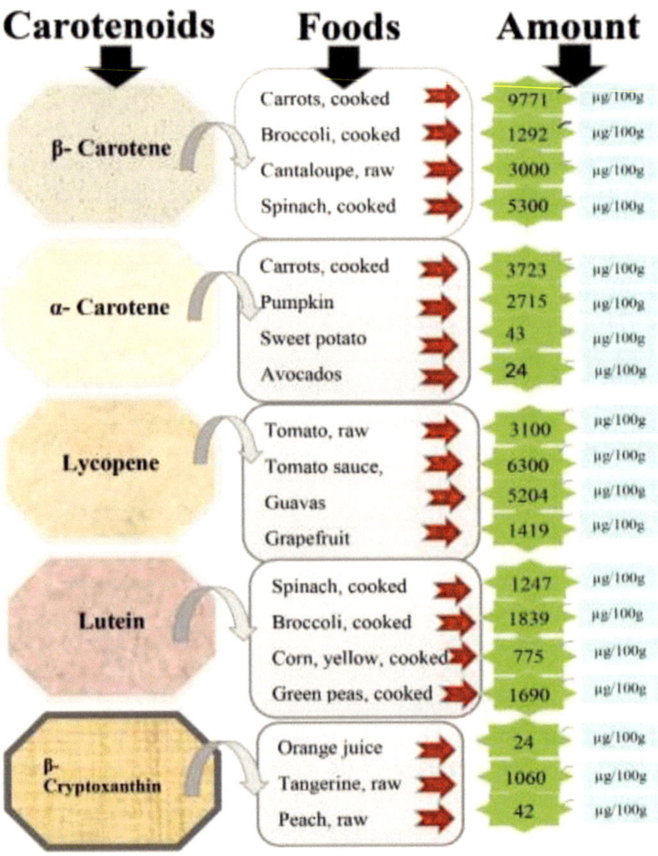

Figure 1.
United States donators of carotenoids rich foods and per capita.

2. Chemistry of carotenoids and their dietary sources

Carotenoids can be found in a variety of foods, although the majority of carotenoids in the diet are derived from strongly colored vegetables, fruits, and juices. Carotenes supplied as food colorings to foods during the process, milk and dairy fat-containing meals, eggs, seafood, and carotenoids provided as food colorings to foods during handling can also supply trace amounts. The principal sources of carotenoids in the United States are shown in **Figure 1**. The data is derived from Median values using current HPLC procedures [20].

B-cryptoxanthin is present in orange fruits, lutein in green leafy vegetables, and lycopene in tomatoes and tomato derivatives, while B-carotene and a-carotene are both found in yellow-orange veggies and fruits. Multicomponent or mixed meals (e.g., soup, stew) generally contain a considerable proportion of carotenoid-rich foods, which is a practical element to address in dietary evaluation [21, 22]. Seasonality may be a key factor of the kind and amount of dietary carotenoids consumed in populations or cultural groups that consume fruits and vegetables in seasonal patterns [23, 24]. Most carotenoids have a polyisoprenoid structure, which means they have a lengthy connected network with the double bonds and are essentially bilaterally symmetrical around the central doubled bond [25]. Multiple carotenoids are generated by cyclizing the end groups and adding oxygen functionalities to the basic structure, which gives them their distinctive hues and antioxidant characteristics. The structure of several carotenoids is shown in **Figure 2**.

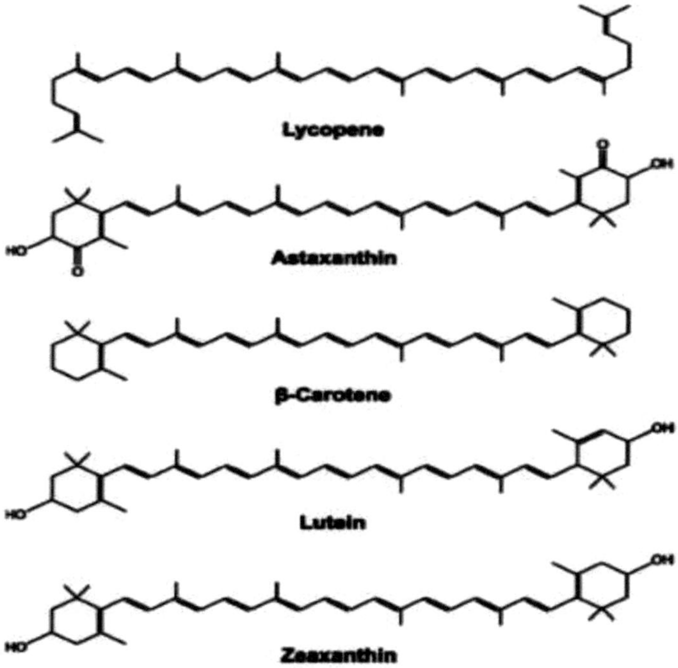

Figure 2.
Chemical structure of common carotenoids.

3. Carotenoids and cardiovascular health

3.1 Lycopene

The most frequent pigment present in human blood is lycopene. That's just a non-cyclic-carotene analog with 11 linked doubled bonds and two distinct doubling bonds arranged in a linear arrangement [25]. This natural pigment is produced by bacteria and plants. Tomatoes are one of the most potent antioxidants, having a respiration activity that is greater than the total beta-carotene and 10 twice that of -tocopherol [26]. This is owing to the high quantity of linked diamines in the product. The adrenals, testicles, liver, and sex organs all contain lycopene [27]. Unlike some other carotenoids, lycopene content in the blood and organs does not correspond well with the total fruit and veggies diet [28].

According to the oxidative hypothesis, preventing LDL from being damaged is the first stage in the production of fibrils and atherosclerosis plaques, which leads to its absorption by monocytes inside the artery wall and the formation of plaque [29]. Oxidative alterations include triglyceride destruction, phospholipid oxidation, and subsequent oxidation of Apolipoprotein B, in addition to unsaturated fatty acids [30].

3.2 Lycopene absorption

Lycopene can mainly be found in its all-trans stereoisomer"s natural form [31]. Lycopene is perhaps the most abundant pigment in blood serum, with a duration of 2–3 days [32].

Fragmentation of the lycopene-rich feed solution, cooking temperature, and the incorporation of lipids as well as other fat compounds, such as other carotenoids, all impact lycopene absorption from food components. Carotenoids, like other lipid-soluble medicines, are digested via a chylomicron-mediated process in the gastrointestinal system [33]. Humans absorb 10–30% of the lycopene they eat in their diet [34, 35]. Sauce, tomato puree, and tomato aqueous extracts capsules all absorb lycopene as well [34, 36]. Lycopene levels are greatest in the testicles, adrenals, prostate, chest, and liver in humans [37, 38]. Lycopene is metabolized and broken down in the tissues. Many oxidizing

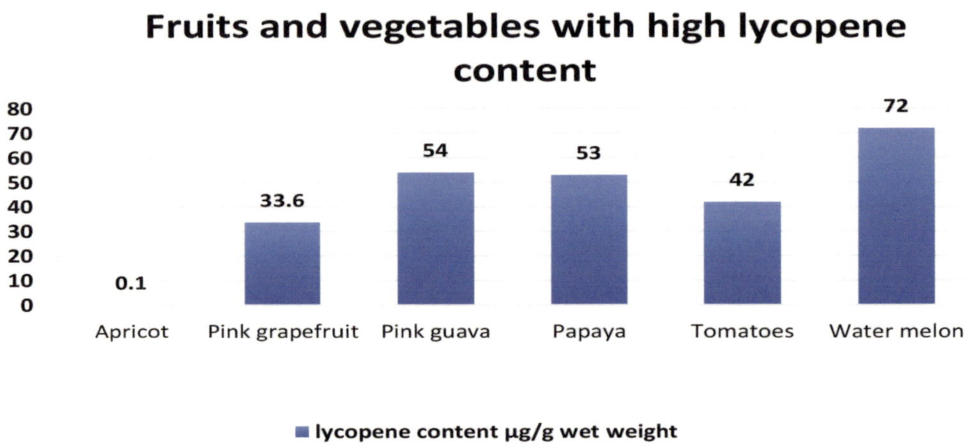

Table 1.
Fruits and vegetables with high lycopene content [37, 40].

lycopene forms, as well as polarized intermediates, have recently been isolated and identified [39]. **Table 1** shows the lycopene content of several foods [37, 40]. According to studies, 10–30% of lycopene taken in the diet is absorbed in the body [41].

3.3 Lycopene and CVD

A lower incidence of cardiovascular disease has been attributed to the Mediterranean diet. Tomatoes, tomato derivatives, lycopene, and other pigments are abundant [42, 43]. In 499 patients with CVD (Mayo cordial infection, strokes, CVD mortality, or revascularization therapies), increased plasma lycopene levels were linked to a decreased risk of cardiovascular disease in the Physicians' Health Study [44]. Lower blood lycopene levels were connected to an increased risk of death in a demographic study comparing Lithuanian and Swedish populations with different heart disease mortality rates [45]. Inflammation is considered to have a part in the development of atherosclerotic disease, which accounts for around 80% of all heart disease cases. In studies, high levels of cytokine production in blood plasma were associated with the onset of cardiac problems [46].

In a research of 139 sick people, oxygenated carotenoids (zeaxanthin, lutein, carotenoids, B-cryptoxanthin, a-carotene, and b-carotene) were found to be reduced in both patient groups (39 with acute illness, 50 with cardiovascular events, and 50 control participants) [47]. In a Japan inhabitants research of 3061 people, there was a link between high blood carotenoids (a-carotene, a-carotene, lycopene, total carotene levels) and a decreased hazard ratio for mortality risk [48]. Upon 60 days of tomato diet intake, a study of CHD (chronic heart disease) patients found a significant improvement in plasma key anti-oxidative enzymes (lipid oxidation rate, dismutase, glutathione peroxidase) compared with control, implying that or more elements of veggies could have medicinal beneficial health effects. In a 3-month study, six healthy guys were given 60 mg of lycopene each day. At the completion of the medication term, their plasma LDL cholesterol level had dropped by 14% [49]. For 1 week, 19 smoke-free healthy people (10 men, 9 women) received lycopene via normal tomatoes and nutraceuticals (20–150 mg/d) in a designed cross nutritional controlled trial [50]. The goal of Thiess and coworkers' randomized clinical experiment was to see how lycopene consumption affected the levels of cardiovascular risk indicators in healthy people. According to the data, taking 10 mg of lycopene every day for 12 weeks did not influence raised blood concentrations. The levels of Apolipoprotein A-I and Apoprotein B-100 were constant. Although the findings were not significant, both the diastolic (DBP) and systolic (SBP) blood pressures were reduced by 3.2 and 0.3 mmHg, respectively [51].

3.4 B-carotene

This group includes the xanthophyll compounds lutein and zeaxanthin. Macula lutea pigments are made from the macula lutea plant's natural dyes. Certain pigments seem to be essential for the physiological function of the eye. They protect against cataracts and macular degeneration caused by aging. These qualities are owed to their antioxidant properties first and foremost [52]. **Figure 1** shows that zeaxanthin has the same composition as lutein and is its derivative. From one of the final b-ion rings, the placement of a double bond changes between the two compounds: zeaxanthin is between C50 and C60, whereas lutein's is between C40 and C5. Leafy foods, along with colorful veggies, are high in lutein. Two of the most prevalent sources are spinach and greens. Lutein can also be present in egg yolks, thanks to the practice of feeding chickens plant-based foods. Corn and red peppers, for example, contain zeaxanthin [53, 54]. Xu and colleagues looked into the efficacy of lutein supplements here on activation of proinflammatory mediators and blood lipids in atherosclerotic subjects. The levels of monocyte chemotaxis protein type 1 in the blood of several individuals who received lutein at a dosage of 20 mg/d for 3 months were decreased (MCP-1). LDL blood cholesterol values were also found to be lower in these individuals [55]. The China Coronary Finding provides evidence that lutein has a protective effect on atherosclerosis. Patients with early stages of atherosclerosis had lower blood lutein concentrations than healthy individuals, according to the study. Plasma lutein content is seen to be negatively linked to the thickness of carotid endothelial tissue (CIMT). The high amount of zeaxanthin in the blood was also shown to be inversely related to right main aorta stiffness and pulse velocity (PWV), both of which are markers of cardiovascular risk [56]. In the 39,876 women who were investigated, there was no link between serum lutein and zeaxanthin contents but also cardiovascular events [57].

3.5 Beta carotene absorption

β-Carotene is a strong fat-soluble nutraceutical that may be found in many fruits and vegetables. β-carotene converts to two molecules of vitamin A, resulting in a higher vitamin A supply [58, 59]. Cardiovascular, cancer, neurological, immunological, rheumatoid arthritis, cataracts, and aging have all been proven to be prevented by β-carotene [60–62]. The effectiveness of tagged β-carotene absorption varies greatly between clinical investigations, ranging from 3 to 80%, but quite often around 10 and 30% [63, 64]. It might be related to β-carotene's varied bioavailability, or it could be owing to the enterocyte's delayed absorption or transit. It's worth noting that the absorption of β-carotene was commonly evaluated after a little meal. In humans, though, our stomach may retain β-carotene from the initial meal for eventual release during the subsequent period [65].

On the other hand, carotenoid binding vehicles may impact carotenoid absorption routes. Blended micelles were most likely separated from the majority of the bolus in the unstirred water of such a glycocalyx region before touching the boundary layer, whereupon carotenoid could be ingested passively or via a transporter-dependent method [66]. Phytofluene, β-carotene, and lutein accumulation are comparable to as well as much bigger than phytoene ingestion in differentiated Caco-2 cell monolayers, albeit lycopene ingestion was the lowest [67, 68]. Uptake efficiency appears to be linked to carotenoid polarity and flexibility in the same manner as bioavailability is. This might be because hydrophilic, pliable pigments have such a stronger attraction for lipids carriers and plasma membrane, resulting in more absorption. According to an IOM report from 2001 [69], the Supplemental and dietary β-carotene absorbing rate ranges from 5 to 26% (spinach) (raw carrots). β-carotene and lycopene are the most abundant carotenoids in human adipocytes, accounting for 20.2 and 18.5% of total carotenoids, respectively, with substantial inter-individual variability [70].

3.6 Beta-carotene and CVD

In a recent meta-analysis of all-cause mortality in 25,468 men and women, the relative risk (RR) for those with the highest vs. lowest blood beta-carotene levels was 0.69 (95% confidence interval: 0.59–0.80). (6137 deaths) [71]. According to the NHANES III study of 16,008 people, some in the top tertile of serum beta-carotene seemed to have a 25% lower risk of mortality (95% CI: 10–37%) than those in the lowest quintile

(4225 deaths) [72]. Many investigations, along with a recent meta-analysis, suggest that circulating beta-carotene and overall mortality are negatively correlated [73–75]. In contrast, a meta-analysis of observational studies found that supplementation with b-carotene raises the odds of cardiovascular mortality from a tiny proportion [76]. Increased nutritional consumption of a-carotene and b-carotene was linked to a reduced risk of CVD mortality in the Zutphen Elderly research [77]. High serum concentrations of a-carotene and b-carotene, lycopene, or carotenoids, according to Japanese population-based follow-up studies, can lower the risk of mortality rates [78, 79].

The development of cardiovascular disorders is undoubtedly aided by peroxidation and chronically low irritation in the cardiovascular system. This pathogenesis of CVD and coronary disease has been related to oxidatively damaged low-density lipoproteins. An injection of such a free radical source that promotes LDL oxidation into foam cells appears to cause thermogenesis. An injection of such a free radical source into foam cells that stimulates LDL oxidation appears to trigger thermogenesis. Antioxidants may prevent cholesterol levels from degradation, lowering the risk of cardiovascular diseases in humans. Because β-carotene and lycopene are mostly found in LDL, they have a significant role in preventing oxidation [80]. The addition of b-carotene to LDL in situ was already found to lower the oxidation sensitivity of LDL [81].

Carotenoids have antioxidant properties and promote lymphocyte proliferation, which would boost immunological activity. The modification of vascular NO bioavailability owing to carotenoids' lowering action is another intriguing technique for explaining how carotenoids assist prevent CVD. In a model of vascular inflammation, high beta-carotene concentrations are connected to a large rise in NO level or absorption, as seen by an increase in cGMP level. In endothelial cells, increased NO release resulted in the enzyme inhibition of NF-kB-dependent binding proteins [82]. Endothelial NO bioavailability is therefore thought to be important to endothelial function and overall vascular health. In a rat model of atherosclerosis, further study reveals that a 9-cis-beta-carotene-rich diet can protect heart disease by lowering non-HDL plasma cholesterol levels, inhibiting liver fibrosis growth and inflammation [83].

3.7 Astaxanthin absorption

Astaxanthin, or 3,3′-dihydroxy-, ′ β-carotene-4,4′-dione, is a red-orange marine carotenoid present inside a wide range of microorganisms and marine animals [61, 62]. Soft gels, capsules, lotions, energy beverages, oils, and extracts containing astaxanthin have already entered the market as nutritional supplements [84]. As for other liposome carotenoids, astaxanthin is considered to go through a complicated digesting and absorption process that includes liberation from food material, transport to a stomach organic phase, creation into micelles under solvation via pancreas hydrolases but also bile acids, transit through the villi, uptake by enterocytes, and inclusion into chylomicrons allowing transportation to the lymphatic vessels and bloodstream [58, 85]. The gastrointestinal system, particularly the duodenum, absorbs almost no carotenoids into enterocytes, and bioavailability refers to the fraction of the ingested dosage retained into micelles [86].

However, because of its weak water solubility and corrosiveness, oral astaxanthin's bioavailability is restricted. The pharmacokinetics of astaxanthin in rats were dose-independent between 100 and 200 mg/kg. Oral astaxanthin intake in the gastrointestinal tract followed a flip-flop pattern, according to Choi et al. [87]. The structure of astaxanthin has a role in its bioavailability. In vitro and rat, experiments demonstrated that a single ingestion of 100 mg mixed isomers resulted in a greater plasma level of cis-astaxanthin, particularly the 13-cis isomer, than diet [88–90].

Osterlie et al. looked at the dispersion of astaxanthin in different lipid fragments and found that 36–64% plasma astaxanthin accumulated in chylomicron-containing very-low-density lipoproteins, with the rest distributed almost evenly between low-density lipoprotein 29% and high-density lipoprotein 24% [90].

3.8 Astaxanthin and CVD

Microalgae, plankton, krill, fish, and other seafood are all members of the xanthophyll family. Microalgae, plankton, krill, fish, and other seafood contain astaxanthin, a red soluble pigment. In the marine environment, it can be found in microalgae, plankton, krill, fish, and some other seafood. It's the pigment that gives salmon and crustaceans their characteristic colors [91]. Even though chronic damage is still a biomarker conducted in a range of diseases, astaxanthin has shown promise in the prevention and treatment of malignancies, inflammatory diseases, metabolic disease, kidney disease, nephropathy, spleen, and digestive diseases, neurodegenerative diseases, and even cardiovascular disease. According to Pashow et al., astaxanthin might help with myocardial injury, oxidation LDL, re-thrombosis following angioplasty, or other cardiac issues including fibrillation. Astaxanthin is a strong anti-oxidant and FR's remover, and a reactive oxygen species (ROS) and nitrogen-oxygen species (NOS) quencher (NOS) [92]. During an eight-week study, Park looked at the effects of astaxanthin supplementation (0, 2, and 8 mg per day) on oxidative stress. People taking 2 mg a day for 8 weeks had a decreased hs-CRP, a primary predictor of heart disease. After 4 weeks of therapy, DNA damage as determined by serum 8-hydroxy-2?-deoxyguanosine was also reduced [93].

3.9 Lutein and zeaxanthin carotene absorption

The xanthophyll pigment astaxanthin (AST) is present in a variety of marine animals and microalgae [28]. Anti-inflammatory and antioxidant capabilities, as well as the ability to improve cardiovascular and immune system health, as well as prevent diabetes and neurological illnesses, are all found in AST [94–98]. In green foods, the lutein-to-zeaxanthin ratio ranges from 12 to 63, with kale having the highest concentration, whereas the ratio in yellow-orange fruits and vegetables is between 0.1 and 1.4 [99]. Dark green algae, that are consumed by fish, are rich in astaxanthin and fucoxanthin. Capsanthin is most commonly found within the pepper. β-Cryptoxanthin is a provitamin A that may be found inside a variety of vegetables, but it's especially abundant in corn, oranges, peaches, papaya, watermelon, and egg yolk [100, 101].

Carotenoids should be digested then delivered into the blood to assert and provide their physical effects. Carotenoids seem to be either lipid-soluble or hydrophilic, indicating they are accessible in fats and immiscible, just like the human digestive tract. When compared to the hydrocarbon carotenoids (α-, β-carotene, and lycopene), lutein and zeaxanthin have hydroxyl groups and are thus polar molecules. To calculate the advantages, a thorough understanding of carotenoid release, absorption, transit, and storage in the eye is required. The quantity and type of dietary fat, that assists in the solvation of releasing carotenoids, and also phospholipids, soluble fiber, and indeed the nature of carotenoids, are all key determinants in lutein and zeaxanthin absorption from food [102–104]. Many phases are engaged in the intake of carotenoids released from food: (i) dispersion inside the stomach colloid so it can be integrated into fat droplets, (ii) followed by translocation to micelles holding bile salts, biliary phospholipids, dietary lipids, as well as other substances.

Food item	Lutein μg/g	Zeaxanthin μg/g
Parsley	64.0–106.5	64.0–106.5
Red pepper	2.5–85.1	5.9–13.5
Corn chips	61.1	92.5
Corn	21.9	10.3
High lutein bread	36.7	3.3
High lutein muffin	26.1	3.7
Durum wheat	5.4	0.5

Table 2.
Lutein and zeaxanthin sources [107–110].

The intestinal cell collects the dissolved carotenoids and distributes them into the blood. In vitro transfer of lutein, zeaxanthin, and β-cryptoxanthin from fruits (orange, kiwi, grapefruit, and sweet potato) was nearly 0%, compared to 19 and 38%, respectively, from spinach and broccoli [105]. The primary carotenoids detected in maize milling fractions are lutein and zeaxanthin, which account for nearly 70% of total carotenoids [106]. **Table 2** lists foods that are high in lutein and zeaxanthin [107–110].

3.10 Zeaxanthin and CVD

With a 40-carbon hydroxylated structure, zeaxanthin is just an oxygenation non-provitamin A carotenoid [111]. The macular lutea, a yellow-colored region of the retina that supports the central vision and includes lutein and zeaxanthin, is a yellow-colored section of the retina that contains lutein and zeaxanthin. Zeaxanthin may protect proteins, lipids, or DNA from oxidative stress via influencing various cellular antioxidant mechanisms, in addition to immediately reducing superoxide radicals. Glutathione is a potent antioxidant found within tissues that defends them from oxidation [112]. Taking supplements with zeaxanthin or a-tocopherol lowers metabolized glutathione (GSSG) levels while raising internalized reduced glutathione (GSH) levels and the GSH/GSSG ratio, particularly during redox balance. By regulating glutathione production and hence glutathione levels, zeaxanthin functions as an antioxidant, either directly or indirectly. As a result, the internal redox state improves in oxidative stress, and susceptibility to H_2O_2-induced cell death decreases [113].

Beta-carotene and zeaxanthin, which are inversely related to right main artery stiffness, pulse speed, and deformability, are implicated in both ocular and cardiovascular health. Both the Beijing and Los Angeles atherosclerosis studies discovered an opposite relation between serum lutein and initial CVD, although subsequent follow-up trials revealed that greater serum zeaxanthin concentrations may defend from early arteriosclerosis [114]. Zeaxanthin may help vascular health, according to these studies.

3.11 Lutein and CVD

With chemically similar formulas, it's an isomer of the carotenoid zeaxanthin. It, like zeaxanthin, is exclusively found in foods like yellow maize, egg yolk, orange juice, honeydew melon, and other fruits, and must be gotten by supplements or diet [113]. The ubiquitous nuclear transcription factor NF-kB, which is implicated in a range of pathogenic reactions, is blocked by lutein [115], as well as the kB inhibitor's degradation (I-kB) [48]. It also has a significant potential to scavenge ROS [116, 117]. NF-kB can begin to migrate into the nucleus when I-kB is released from the NF-kB complex by lutein, reducing inducible transcription of genes and the activation of cytokines markers such cytokines, chemokines, and iNOS [118]. Lutein inhibits the production of TNF-alpha, interleukin 6 (IL-6), prostaglandin 2 (PGE-2), monocyte chemotactic protein 1 (MCP-1), and macrophage inflammatory protein 2 (MIP-2) [119].

According to this study, plasmatic lutein shows a negative correlation with oxidative stress, implying that it has significant oxidative and anti-inflammatory effects on aortic tissue, potentially preventing atherosclerotic [120]. Several studies have found that lutein levels in individuals with atherosclerosis were considerably lower than in normal and that they were indirectly correlated to arterial tightness [121]. The cardiac and blood vessel preventive actions of lutein have also been connected to the management of hypertension. A rise in systolic blood pressure and unintentional hypertension was often negatively proportional to a larger amount of this carotenoid. Some with greater lutein values had lower serum blood pressure but a decreased risk of future hypertension, independent of whether or not they smoke [122]. By lowering peroxidation and myocyte apoptosis, lutein prevents the myocardium from ischemia damage [123]. By avoiding contractile dysfunction, limiting myocardial damage may lower CAD morbidity and mortality [124].

3.12 Oxidative stress and antioxidants activity

The onset and evolution of a range of disorders, including cardiovascular issues, have been related to oxidative stress. ROS are important biological variables that can influence a wide range of physiological and disease-related conditions [125]. Cancer, reactive arthritis, osteoarthritis, aging, neurological, and cardiac illnesses are all connected to oxidative stress. Given the evidence linking oxidative stress to a wide range of human illnesses, measuring oxidative stress biomarkers is critical for assessing health and detecting the onset of oxidative stress-related disorders [126]. Hypercholesterolemia is also a disease that is closely connected to peroxidation. FH individuals reported greater levels of reactive oxygen species than normolipidemic patients, as per an inter observational study involving 132 individuals with high cholesterol (FH). MDA concentration seems to be much higher in FH, suggesting a higher oxidative stress state, according to the International Federation of Clinical Chemists (IFCC) standard range (>1.24 g/L) [127]. Various demographic studies have examined the association between higher nutritional carotene intake and thus the environment's effects on cardiovascular disease prevention [128–131].

Circulating carotenoid concentrations, for example, have been associated with inflammatory markers, increased lipid peroxidation, and vascular dysfunction, that has all been connected to CVD [132–134]. Secondly, pigments and minerals have a phytonutrient-like impact on endothelial dysfunction and irritation, decreasing the risk of atherosclerosis [135]. The finding of a link between carotene, peroxidation, and inflammation has been aided by several in vitro studies, notably those that used subsystem [136]. Carotenoids exhibit anti-oxidant and anti-inflammatory properties in vascular cells, as shown in **Figure 3**.

Nitrogen oxide may combine with O^{2-} to generate peroxynitrite (ONOO-) under oxidative conditions, resulting in decreased NO bioavailability, vascular dysfunction, increased lipid oxidation, and chronic inflammatory responses. Nitrogen oxide may combine with oxygen to generate peroxynitrite (ONOO-) during oxidative

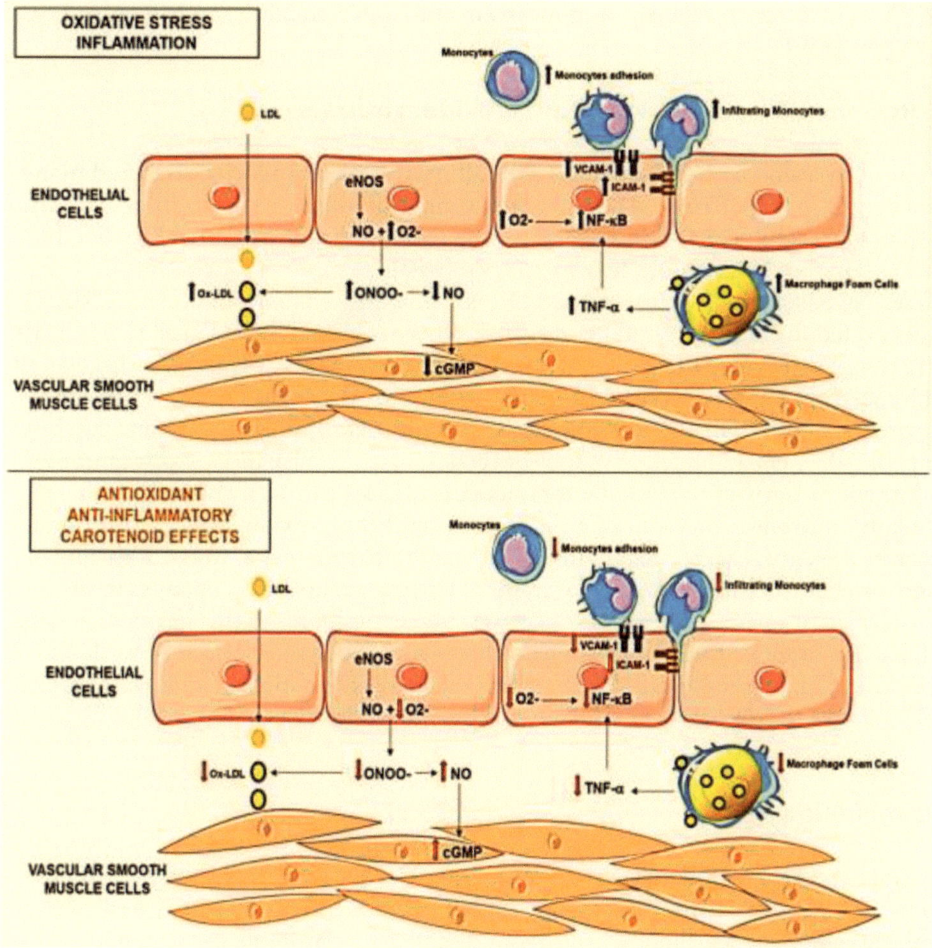

Figure 3.
Carotenoids have a beneficial effect on endothelial dysfunction and overall vascular health [137].

conditions, resulting in decreased NO bioavailability, vascular dysfunction, increased lipid oxidation, and chronic inflammatory responses. All of these actions create a negative cycle, and the antioxidant and anti-inflammatory capabilities of carotenoids may be harmed as a result. TNF-, tumor necrotic lesions factor-alpha; NF-B, nuclear factor kappa-light-chain-enhancer of activated B cells; ICAM-1, intercellular adhesion molecule 1; VCAM-1, vascular cell adhesion molecule 1; TNF-, tumor necrosis factor-alpha; NF-B, nuclear factor kappa-light-chain-enhancer of activated B cells; ICAM-1, intercellular adhesion molecule 1; TNF-, tumor necrosis factor-alpha; NF-B, nuclear factor kappa-light-chain-enhancer of activated B cells; ICAM-1, intercellular adhesion molecule 1; eNOS, endothelial nitric oxide synthase; NO, nitric oxide; O_2-, superoxide anion; ONOO-, peroxynitrite; eNOS, endothelial nitric oxide synth cGM.

Antioxidants are chemicals that prevent or restrict oxidative damage by inhibiting the action of reactive oxygen species. Intrinsic antioxidant components present in body cells include chronic damage, catalase, and glutathione peroxidase. Antioxidants found in foods include vitamin C, vitamin C, polyphenols, and carotenoids [138]. To help avoid chronic diseases like cancer and cardiovascular disease, current dietary guidelines recommend consuming more antioxidant-rich plant foods like fruits and vegetables [139].

3.13 ROS and RNS production by nitric oxide synthases

The conversion of L-arginine to L-citrulline and nitric oxide is catalyzed by nitric oxide synthases (NOS)., but they can create superoxide under uncoupling conditions:

NOS + L-arginine + O_2 + NADPH – → NO + citrulline + NADP+ NOS(Fe(II) heme) + O_2 – → NOS(Fe(III)heme) + O_2 •– [140].

Two NOS isoforms, neuronal NOS (NOS1) and endothelial NOS (eNOS, NOS3), are generated in cardiomyocytes constantly, whereas inducible NOS (NOS2) is lacking in the healthy heart but can be triggered by pro-oxidants [141]. It was discovered that hypertrophied myocytes had a higher amount of iNOS [142]. Because NOSs may produce both RNS and ROS, their effects on the cardiovascular system can be complex—they can enhance or reduce heart damage. Because nitric oxide is an EDRF (endothelium-derived relaxing factor), its effects must primarily benefit the heart. The diffusion-controlled interaction of nitric oxide with superoxide, on the other hand, produces the very reactive peroxynitrite. To avoid heart damage, the equilibrium of superoxide/nitric oxide must be maintained. During pathological changes in the heart, the interplay of major enzymatic ROS generators contributes to this balance. In dogs with pacing-induced heart failure, NO synthases and xanthine oxidase was shown to be important in the modulation of myocardial mechanical efficiency, and overexpression of XO relative to NOS contributed to mechanoenergetic uncoupling [143].

4. Conclusion

Fruits and vegetables are rich in carotenoids. Carotenoids have long been regarded to be beneficial to one's health. Nearly 700 carotenoids have been discovered. The most regularly referenced carotenoids in this chapter were a-carotene, b-carotene, lutein, lycopene, and zeaxanthin. Their absorption, transportation, needs, and chemistry were all discussed. Cardiovascular diseases are a significant public health issue. Carotenoids-rich meals may help to reduce the progression of coronary heart disease, according to the study reviewed in this chapter. Oxidative stress is responsible for a wide range of degenerative diseases, including cardiovascular issues. The pathogenesis of CVD is heavily influenced by oxidative stress. We looked at the significance of carotenoids in endothelial function and vascular health in general in this chapter. We also discussed how carotenoids may be obtained from a variety of fruits and vegetables. The etiology of atherosclerosis is aggravated by oxidative stress. Throughout this chapter, we looked at the significance of carotenoids in endothelial function and vascular health in general.

Acknowledgements

We thank the digital library GCUF for providing access to the publication.

Conflict of interest

There is no conflict of interest as declared by all authors.

Author details

Arslan Ahmad[1*], Sakhawat Riaz[1], Muhammad Shahzaib Nadeem[1], Umber Mubeen[1] and Khadija Maham[2]

1 Department of Home Economics, Government College University Faisalabad, Faisalabad, Pakistan

2 Department of Nutritional Science, Government College University Faisalabad, Faisalabad, Pakistan

*Address all correspondence to: ahmadarslan784@gmail.com

References

[1] Krinsky NI. Actions of carotenoids in biological systems. Annual Review of Nutrition. 1993;**13**:561-587

[2] Crews HA, Link G, Andersen R, Bradesco V, Holst B, Maiani G, et al. A critical assessment of some biomarker approaches linked with dietary intake. The British Journal of Nutrition. 2001;**86**:5-35

[3] Olson JA. In: Porter JW, Spurgeon SL, editors. Formation and Function of Vitamin a. Biosynthesis of Isoprenoid Compounds. Vol. 2. New York: Wiley & Sons; 1983. pp. 371-412

[4] Parker RS. Carotenoids in human blood and tissues. The Journal of Nutrition. 1989;**119**:101-104

[5] Khachik F, Beecher GR, Whittaker NF. Separation, identification, and quantification of the major carotenoid and chlorophyll constituents in extracts of several green vegetables by liquid chromatography. Journal of Agriculture Food Chemistry. 1986;**34**:603-661

[6] Mathews-Roth MM, Gulbrandsen CL. Transport of beta-carotene in serum of individuals with carotenemia. Clinical Chemistry. 1974;**20**:1578-1579

[7] Olson JA. The provitamin a function of carotenoids: The conversion of beta-carotene into vitamin a. The Journal of Nutrition. 1989;**119**:105-108

[8] Mangels AR, Holden JM, Beecher GR, Forman MR, Lanza E. Carotenoid contents of fruits and vegetables: An evaluation of analytical data. Journal of the American Dietetic Association. 1993;**93**:284-296

[9] Johnson EJ. The role of carotenoids in human health. Nutrition in Clinical Care. 2002;**5**(2):47-49

[10] Agarwal S, Rao AV. Carotenoids and chronic diseases. Drug Metabolism and Drug Interactions. 2000;**17**(1-4):189-210

[11] Leermakers ETM, Darweesh SKL, Baena CP, Moreira EM, Van Lent DM, Tielemans MJ, et al. The effects of lutein on cardiometabolic health across the life course: A systematic review and meta-analysis. The American Journal of Clinical Nutrition. 2016;**103**(2):481-494. DOI: 10.3945/ajcn.115.120931. Retrieved from

[12] Ravikrishnan R, Rusia S, Ilamurugan G, Salunkhe U, Deshpande J, Shankaranarayanan J, et al. Safety assessment of lutein and zeaxanthin (LutemaxTM 2020): Subchronic toxicity and mutagenicity studies. Food and Chemical Toxicology. 2011;**49**(11):2841-2848. DOI: 10.1016/j. fct.2011.08.011. Retrieved from

[13] Vijay K, Sowmya PRR, Arathi BP, Shilpa S, Shwetha HJ, Raju M, et al. Low-dose doxorubicin with carotenoids selectively alters redox status and upregulates oxidative stress-mediated apoptosis in breast cancer cells. Food and Chemical Toxicology. 2018;**118**:675-690. DOI: 10.1016/j. fct.2018.06.027. Retrieved from

[14] Zaccara G, Lattanzi S, Cincotta M, Russo E. Drug treatments in patients with cardiac diseases and epilepsy. Acta Neurologica Scandinavica. 2020;**142**(1):37-49. DOI: 10.1111/ ane.13249. Retrieved from

[15] Albert CM, Deo R. Epidemiology and genetics of sudden cardiac death. Circulation. 2012;**125**(4):620-637 Retrieved from. http://circ. ahajournals.org/cgi/doi/1 0.1161/CIR CULATIONAHA.111.023838%5Cn

papers3://publication/DOI/10.1161/CIRCULATIONAHA.111.023838

[16] Maria AG, Graziano R, Nicolantonio DO. Carotenoids: Potential allies of cardiovascular health? Food and Nutrition Research. 2015;**59**(1):26762. DOI: 10.3402/fnr.v59.26762

[17] Ribeiro D, Freitas M, Silva AMS, Carvalho F, Fernandes E. Antioxidant and pro-oxidant activities of carotenoids and their oxidation products. Food and Chemical Toxicology. 2018;**120**:681-699. DOI: 10.1016/j.fct.2018.07.060. Retrieved from

[18] Chung RWS, Leanderson P, Lundberg AK, Jonasson L. Lutein exerts anti-inflammatory effects in patients with coronary artery disease. Atherosclerosis. 2017b;**262**:87-93. DOI: 10.1016/j.atherosclerosis.2017.05.008. Retrieved from

[19] Heller FR, Descamps O, Hondekijn JC. LDL oxidation: Therapeutic perspectives. Atherosclerosis. 1998;**137**(Suppl):S25-S31

[20] Mangels AR, Holden JM, Beecher GR, Forman MR, Lanza E. Carotenoid content of fruits and vegetables: An evaluation of analytic data. Journal of the American Dietetic Association. 1993;**93**:284-296

[21] Chug-Ahuja JK, Holden JM, Forman MR, Mangels AR, Beecher GR, Lanza E. The development and application of a carotenoid database for fruits, vegetables, and selected multicomponent foods. Journal of the American Dietetic Association. 1993;**93**:318-323

[22] Hankin JE, Le Marchand L, Kolonel LN, Wilkens LR. Assessment of carotenoid intakes in humans. Annals NY Academy Science. 1993;**691**:68-75

[23] Olmedilla B, Granado F, Blanco I, Rojas-Hidalgo E. Seasonal and sex-related variations in six serum carotenoids, retinol, and alpha-tocopherol. The American Journal of Clinical Nutrition. 1994;**60**:106-l 10

[24] Rautalahti M, Alhanes D, Huakka J, Roos E, Gref CG, Virtamo J. Seasonal variation of serum concentrations of beta-carotene and alpha-tocopherol. The American Journal of Clinical Nutrition. 1993;**57**:551-556

[25] Britton G. Structure and properties of carotenoids about function. The FASEB Journal. 1995;**9**:1551-1558

[26] Miller NJ, Sampson J, Candeias LP, et al. Antioxidant activities of carotenes and xanthophylls. FEBS Letters. 1996;**384**:240-246

[27] Stahl W, Schwarz W, Sundquist AR, et al. Cis-trans isomers of lycopene and beta-carotene in human serum and tissues. Archives of Biochemistry and Biophysics. 1992;**294**:173-177

[28] Michaud DS, Giovannucci EL, Ascherio A, et al. Associations of plasma carotenoid concentrations and dietary intake of specific carotenoids in samples of two prospective cohort studies using a new carotenoid database. Cancer Epidemiology, Biomarkers & Prevention. 1998;**7**:283-290

[29] Witztum JL. The oxidation hypothesis of atherosclerosis. Lancet. 1994;**344**:793-795

[30] Witztum JL. Role of oxidized low-density lipoprotein in atherogenesis. British Heart Journal. 1993;**69**:S12-S18

[31] Clinton SK. Lycopene: Chemistry, biology, and implications for human health and disease. Nutrition Reviews. 1998;**1**:35-51

[32] Stahl W, Sies H. Lycopene: A biologically important carotenoid for

humans? Archives of Biochemistry and Biophysics. 1996;**336**:1-9

[33] Parker RS. Absorption, metabolism, and transport of carotenoids. The FASEB Journal. 1996;**10**:542-551

[34] Gartner C, Stahl W, Sies H. Lycopene is more bioavailable from tomato ¨ paste than from fresh tomatoes. The American Journal of Clinical Nutrition. 1997;**66**:116-122

[35] Rao AV, Agarwal S. Role of lycopene as antioxidant carotenoid in the prevention of chronic diseases: A review. Nutrition Research. 1999;**19**:305-323

[36] Rao AV, Agarwal S. Bioavailability and antioxidant properties of lycopene from tomato products. Nutrition and Cancer. 1998;**31**:199-203

[37] Rao AV, Mira MR, Rao LG. Lycopene. Advances in Food and Nutrition Research. 2006;**51**:99-164

[38] Jain CK, Agarwal S, Rao AV. The effect of dietary lycopene on bioavailability, tissue distribution, in-vivo antioxidant properties, and colonic preneoplasia in rats. Nutrition Research. 1999;**19**:1383-1391

[39] Khachik F, Carvallo L, Bernstein PS, Muir GJ, Zhao DY, Katz NB. Chemistry, distribution, and metabolism of tomato carotenoids and their impact on human health. Experimental Biology and Medicine. 2002;**227**(10):845-851

[40] Rao AV, Rao LG. Lycopene and human health. Current Topical Nutrition Research. 2004;**2**:127-123

[41] Palozza P, Catalano A, Simone R. E, Mele M, C, Cittadini a: Effect of lycopene and tomato products on cholesterol metabolism. Annals of Nutrition & Metabolism. 2012;**61**:126-134. DOI: 10.1159/000342077

[42] Fuhramn B, Elis A, Aviram M. Hypocholesterolemic effect of lycopene and ß-carotene is related to the suppression of cholesterol synthesis and augmentation of LDL receptor activity in macrophage. Biochemical and Biophysical Research Communications. 1997;**233**:658-662

[43] Agarwal S, Rao AV. Tomato lycopene and low-density lipoprotein oxidation: A human dietary intervention study. Lipids. 1998;**33**:981-984

[44] Sesso HD, Buring JE, Norkus EP, et al. Plasma lycopene, other carotenoids, retinol and the risk of cardiovascular disease in men. The American Journal of Clinical Nutrition. 2005;**81**:990-997

[45] Kristenson M, Zieden B, Kucinskiene Z, et al. Antioxidant state and mortality from coronary heart disease in Lithuanian and Swedish men: Concomitant cross-sectional study of men aged 50. BMJ. 1997;**314**:629-633

[46] Mann DL et al. Targeted Anticytokine therapy and the failing heart. The American Journal of Cardiology. 2005;**95**(11, Suppl. 1):9-16

[47] Lidebjer C, Leanderson P, Ernerudh J, et al. Low plasma levels of oxygenated carotenoids in patients with coronary artery disease. Nutrition, Metabolism, and Cardiovascular Diseases. 2007;**17**:448-456

[48] Ito Y, Kurata M, Suzuki K, Hamajima N, et al. Cardiovascular disease mortality and serum carotenoid levels: A Japanese population-based follow-up study. Journal of Epidemiology. 2006;**16**:154-160

[49] Bose KS, Agrawal BK. Effect of lycopene from cooked tomatoes on serum antioxidant enzymes, lipid peroxidation rate, and lipide profile in coronary heart disease. Singapore Medical Journal. 2007;**48**:415-420

[50] Rao AV, Agarwal S. Bioavailability and in vivo antioxidant properties of lycopene from tomato products and their possible role in the prevention of cancer. Nutrition and Cancer. 1998;**31**:199-203

[51] Thies F, Masson LF, Rudd A, Vaughan N, Tsang C, Brittenden J, et al. Effect of a tomato-rich diet on markers of cardiovascular disease risk in moderately overweight, disease-free, middle-aged adults: A randomized controlled trial. The American Journal of Clinical Nutrition. 2012;**95**(5):1013-1022. DOI: 10.3945/ajcn.111.026286

[52] Leermakers ETM, Moreira EM, Kiefte-de Jong JC, Darweesh SKL, Visser T, Voortman T, et al. Effects of choline on health across the life course: A systematic review. Nutrition Reviews. 2015;**73**(8):500-522. DOI: 10.1093/nutrit/nuv010

[53] Holden JM, Eldridge AL, Beecher GR, Buzzard IM, Bhagwat S, Davis CS, et al. Carotenoid content of U.S. foods: An update of the database. Journal of Food Composition and Analysis. 1999;**12**(3):169-196. DOI: 10.1006/jfca.1999.0827

[54] Abdel-Aal E-SM, Akhtar H, Zaheer K, Ali R. Dietary sources of lutein and zeaxanthin carotenoids and their role in eye health. Nutrients. 2013;**5**: 1169-1185. DOI: 10.3390/nu5041169

[55] Zou XR, Xiao ZYX, Huang YM, Wang X, Lin XM. Effects of lutein supplement on serum inflammatory cytokines, ApoE and lipid profiles in early atherosclerosis population. Journal of Atherosclerosis and Thrombosis. 2013;**20**(2):170-177

[56] Zou Z, Xu X, Huang Y, Xiao X, Ma L, Sun T, et al. High serum level of lutein may be protective against early atherosclerosis. The Beijing Atherosclerosis Study. 2011;**219**(2):789-793

[57] Sesso HD, Buring JE, Norkus EP, Gaziano JM. Plasma lycopene and the risk of cardiovascular disease in women. Journal of the American College of Cardiology. 2012;**39**:147

[58] Donhowe GE, Kong F. Beta-carotene: Digestion, microencapsulation, and In vitro bioavailability. Food and Bioprocess Technology. 2014;**7**:338-354. DOI: 10.1007/s11947-013-1244-z

[59] Loveday S, Singh H. Recent advances in technologies for vitamin a protection in foods. Trends in Food Science and Technology. 2008;**19**:657-668

[60] Yi J, Lam T, Yokoyama W, Cheng L, Zhong F. Beta-carotene encapsulated in food protein nanoparticles reduces peroxyl radical oxidation in Caco-2 cells. Food Hydrocolloids. 2015;**43**

[61] de Freitas Zômpero RH, López-Rubio A, de Pinho SC, Lagaron JM, de la Torre LG. Hybrid encapsulation structures based on β-carotene-loaded nanoliposomes within electrospun fibers. Colloids Surfaces B Biointerfaces [Internet]. 2015;**134**:475-482 Available from: http://www.sciencedirect.com/science/article/pii/S0927776515001496

[62] Jain A, Thakur D, Ghoshal G, Katare OP, Shivhare US. Characterization of microcapsulated β-carotene formed by complex Coacervation using Casein & gum Tragacanth [internet]. International Journal of Biological Macromolecules. Elsevier B.V. 2016. DOI: 10.1016/j.ijbiomac.2016.01.117

[63] van Lieshout M, West CE, van Breemen RB. Isotopic tracer techniques for studying the bioavailability and bioefficacy of dietary carotenoids, particularly beta-carotene, in humans: A

review. The American Journal of Clinical Nutrition. 2003;**77**:12-28 [CrossRef]

[64] Van Loo-Bouwman CA, Naber TH, van Breemen RB, Zhu D, Dicke H, Siebelink E, et al. Vitamin an equivalency and apparent absorption of beta-carotene in ileostomy subjects using a dual-isotope dilution technique. The British Journal of Nutrition. 2010;**103**:1836-1843 [CrossRef]

[65] Borel P, Grolier P, Mekki N, Boirie Y, Rochette Y, Le Roy B, et al. Low and high responders to pharmacological doses of beta-carotene: Proportion in the population, mechanisms involved and consequences on beta-carotene metabolism. Journal of Lipid Research. 1998;**39**:2250-2260

[66] Phan CT, Tso P. Intestinal lipid absorption and transport. Frontiers in Bioscience. 2001;**6**:D299-D319 [CrossRef] [PubMed]

[67] During A, Hussain MM, Morel DW, Harrison EH. Carotenoid uptake and secretion by caco-2 cells: Beta-carotene isomer selectivity and carotenoid interactions. Journal of Lipid Research. 2002;**43**:1086-1095 [CrossRef]

[68] Mapelli-Brahm P, Desmarchelier C, Margier M, Reboul E, Melendez Martinez AJ, Borel P. Phytoene and phytofluene isolated from a tomato extract are readily incorporated in mixed micelles and absorbed by caco-2 cells, as compared to lycopene, and sr-bi is involved in their cellular uptake. Molecular Nutrition & Food Research. 2018;**62**:e1800703 [CrossRef] [PubMed]

[69] Institute of Medicine. Dietary Reference Intakes for Vitamin a, Vitamin K, Arsenic, Boron, Chromium, Copper, Iodine, Iron, Manganese, Molybdenum, Nickel, Silicon, Vanadium, and Zinc. Washington, DC: National Academies Press; 2001

[70] Parker RS. Carotenoid and tocopherol composition of human adipose tissue. The American Journal of Clinical Nutrition. 1988;**47**:33-36

[71] Zhao LG, Zhang QL, Zheng JL, Li HL, Zhang W, Tang WG, et al. Dietary, circulating betacarotene and risk of all-cause mortality: A meta-analysis from prospective studies. Scientific Reports. 2016;**6**:26983

[72] Goyal A, Terry MB, Siegel AB. Serum antioxidant nutrients, vitamin a, and mortality in u.S. adults. Cancer Epidemiology, Biomarkers & Prevention. 2013;**22**:2202-2211

[73] Hashim D, Gaughan D, Boffetta P, Lucchini RG. Baseline serum beta-carotene concentration and mortality among long-term asbestos-exposed insulators. Cancer Epidemiology, Biomarkers & Prevention. 2015;**24**:555-560

[74] Bates CJ, Hamer M, Mishra GD. Redox-modulatory vitamins and minerals that prospectively predict mortality in older British people: The national diet and nutrition survey of people aged 65 years and over. The British Journal of Nutrition. 2011;**105**:123-132

[75] Kilander L, Berglund L, Boberg M, Vessby B, Lithell H. Education, lifestyle factors and mortality from cardiovascular disease and cancer. A 25-year follow-up of Swedish 50-year-old men. International Journal of Epidemiology. 2001;**30**:1119-1126

[76] Vivekananthan DP, Penn MS, Sapp SK, Hsu A, Topol EJ. Use of antioxidant vitamins for the prevention of cardiovascular disease: A meta-analysis of randomized trials. Lancet. 2003;**361**:2017e23

[77] Buijsse B, Feskens EJ, Kwape L, Kok FJ, Kromhout D. Both alpha- and

beta-carotene, but not tocopherols and vitamin C, are inversely related to 15-year cardiovascular mortality in Dutch elderly men. The Journal of Nutrition. 2008;**138**:344e50

[78] Ito Y, Kurata M, Suzuki K, Hamajima N, Hishida H, Aoki K. Cardiovascular disease mortality and serum carotenoid levels: A Japanese population-based follow-up study. Journal of Epidemiology. 2006;**16**:154e60

[79] Ito Y, Suzuki K, Ishii J, Hishida H, Tamakoshi A, Hamajima N, et al. A population-based follow-up study on mortality from cancer or cardiovascular disease and serum carotenoids, retinol and tocopherols in Japanese inhabitants. Asian Pacific Journal of Cancer Prevention. 2006;**7**:533e46

[80] Tapiero H, Townsend DM, Tew KD. Dossier: Influence of alcohol consumption and smoking habits on human health, the role of carotenoids in the prevention of human pathologies. Biomed & Pharmacotherapy. 2004;**58**: 100-110

[81] Jialal I, Norkus EP, Cristol L, Grundy SM. Beta-carotene inhibits the oxidative modification of low-density lipoprotein. Biochem Biophys Acta. 1991;**1086**:1348

[82] Aizawa T, Wei H, Miano JM, Abe J. Role of PDE-3 in NO/cGMP-mediated anti-inflammatory effects in vascular smooth muscle cells. Circulation Research. 2003;**93**:40613

[83] Harari A, Harats D, Marko D, Cohen H, Barshack I, Kamari Y, et al. A 9-cis b-Caroteneenriched diet inhibits atherogenesis and fatty liver formation in LDL-R knockout mice. Journal of Nutrition Diseases. 2008;**138**:192330

[84] Bogacz-Radomska L, Harasym J, Piwowar A. Commercialization aspects of carotenoids. Carotenoids: Properties, Processing, and Applications. 2020;**10**:327-357. DOI: 10.1016/B978-0-12-817067-0.00010-5

[85] Saini RK, Nile SH, Park SW. Carotenoids from fruits and vegetables: Chemistry, analysis, occurrence, bioavailability, and biological activities. Food Research International (Ottawa, Ont.). 2015;**76**(Pt 3):735-750. DOI: 10.1016/j.foodres.2015.07.047

[86] Sy C, Gleize B, Dangles O, Landrier JF, Veyrat CC, Borel P. Effects of physicochemical properties of carotenoids on their bioaccessibility, intestinal cell uptake, and blood and tissue concentrations. Molecular Nutrition & Food Research. 2012;**56**(9): 1385-1397. DOI: 10.1002/mnfr.201200041

[87] Choi HD, Youn YK, Shin WG. Positive effects of astaxanthin on lipid profiles and oxidative stress in overweight subjects. Plant Foods for Human Nutrition (Dordrecht, Netherlands). 2011;**66**(4):363-369. DOI: 10.1007/s11130-011-0258-9

[88] Honda M, Murakami K, Osawa Y, Kawashima Y, Hirasawa K, Kuroda I. Z-isomers of astaxanthin exhibit greater bioavailability and tissue accumulation efficiency than the all-E-isomer. Journal of Agricultural and Food Chemistry. 2021;**69**(11):3489-3495. DOI: 10.1021/acs.jafc.1c00087

[89] Yang C, Zhang H, Liu RH, Zhu HH, Zhang LF, Tsao R. Bioaccessibility, cellular uptake, and transport of astaxanthin isomers and their antioxidative effects in human intestinal epithelial Caco-2 cells. Journal of Agricultural and Food Chemistry. 2017;**65**(47):10223-10232. DOI: 10.1021/acs.jafc.7b04254

[90] Osterlie M, Bjerkeng B, Liaaen-Jensen S. Plasma appearance and

distribution of plasma lipoproteins of men after single-dose administration of astaxanthin. Journal of Nutritional Biochemistry. 2000;**11**(10):482-490. DOI: 10.1016/S0955-2863(00)00104-2

[91] Schweigert F. Metabolism of Carotenoids in Mammals. Basel, Switzerland: Birkhauser Verlag; 1998

[92] Nakano M, Onodera A, Saito E, Tanabe M, Yajima K, Takahashi J, et al. Effect of astaxanthin in combination with alpha-tocopherol or ascorbic acid against oxidative damage in diabetic ODS rats. Journal of Nutritional Science and Vitaminology. 2008;**54**:32934

[93] Park JS, Chyun JH, Kim YK, Line LL, Chew BP. Astaxanthin decreased oxidative stress and inflammation and enhanced immune response in humans. Nutrition and Metabolism. 2010;7:18

[94] Yasui Y, Hosokawa M, Mikami N, Miyashita K, Tanaka T. Dietary astaxanthin inhibits colitis and colitisassociated colon carcinogenesis in mice via modulation of the inflammatory cytokines. Chemico-Biological Interactions. 2011;**193**(1):79-87

[95] Yang Y, Kim B, Lee JY. Astaxanthin structure, metabolism, and health benefits. Journal of Human Nutrition & Food Science. 2013;**1**:1003

[96] Grimmig B, Kim SH, Nash K, Bickford PC, Shytle RD. Neuroprotective mechanisms of astaxanthin: A potential therapeutic role in preserving cognitive function in age and neurodegeneration. Geroscience. 2017;**39**(1):19-32

[97] Visioli F, Artaria C. Astaxanthin in cardiovascular health and disease: Mechanisms of action, therapeutic merits, and knowledge gaps. Food & Function. 2017;**8**(1):39-63

[98] Guan L, Liu J, Yu H, Tian H, Wu G, Liu B, et al. Water-dispersible astaxanthin-rich nanopowder: Preparation, oral safety, and antioxidant activity in vivo. Food & Function. 2019;**10**(3):1386-1397

[99] Humphries JM, Khachik F. Distribution of lutein, zeaxanthin, and related geometrical isomers in fruits, vegetable, wheat, and pasta products. Journal of Agricultural and Food Chemistry. 2003;**51**:1322-1327

[100] Chandrika UG, Jansz ER, Wickranasinghe SMDN, Warnasuriya ND. Carotenoids in yellow and red-fleshed papaya (Carcia papaya L). Journal of the Science of Food and Agriculture. 2003;**83**:1279-1282

[101] United States Department of Agriculture. Nutritional data laboratory home page. USDA Nutritional database for standard reference release 22. 2009. Available online: http://www.ars.usda.gov/Services/docs.htm?docid=20960 [Accessed November 24, 2012].

[102] Van Het Hof KH, Weststrate JA, Hautvast JG. Dietary factors that affect the bioavailability of carotenoids. Nutrition Research. 1999;**130**:503-506

[103] Castenmiller JJ, West CE. Bioavailability and bioconversion of carotenoids. Annual Review of Nutrition. 1998;**18**:19-38

[104] Bohn T. Bioavailability of non-provitamin a carotenoids. Current Nutrition & Food Science. 2008;**4**:240-258

[105] O'Connell OF, Ryan L, O'Brien NB. Xanthophyll carotenoids are more bioaccessible from fruits than dark green vegetables. Nutrition Research. 2007;**27**:258-264

[106] Kean EG, Hamaker BR, Ferruzzi MG. Carotenoids

bioaccessibility from whole grain and degermed maize meal products. Journal of Agricultural and Food Chemistry. 2008;**56**:9918-9926

[107] Maiani G, Periago Caston MJ, Catasta G, Toti E, Cambrodon IG, Bysted A, et al. Carotenoids: Actual knowledge on food sources, intakes, stability and bioavailability and their protective role in humans. Molecular Nutrition & Food Research. 2009;**53**:S194-S218

[108] Abdel-Aal E-SM, Young JC, Akhtar H, Rabalski I. Stability of lutein in wholegrain bakery products naturally high in lutein or fortified with free lutein. Journal of Agricultural and Food Chemistry. 2010;**58**:10109-10117

[109] De La Parra C, Saldivar SOS, Lui RH. Effect of processing on the phytochemical profiles and antioxidant activity of corn for production of masa, tortillas, and tortilla chips. Journal of Agricultural and Food Chemistry. 2007;**55**:4177-4183

[110] Abdel-Aal E-SM, Young JC, Rabalski I, Frégeau-Reid J, Hucl P. Identification, and quantification of seed carotenoids in selected wheat species. Journal of Agricultural and Food Chemistry. 2007;**55**:787-794

[111] Pfander H. Carotenoids: an over-view. Methods in Enzymology. 1992;**213**:313

[112] Giblin FJ. Glutathione: a vital lens antioxidant. Journal of Ocular Pharmacology and Therapeutics. 2000;**16**:12135

[113] Gao S, Qin T, Liu Z, Caceres MA, Ronchi CF, Chen CYO, et al. Lutein and zeaxanthin supplementation reduces H2O2- induced oxidative damage in human lens epithelial cells. Molecular Vision. 2011;**17**:318090

[114] Dwyer JH, Paul-Labrador MJ, Fan J. Progression of carotid intima-media thickness and plasma antioxidants: The Los Angeles atherosclerosis study. Arteriosclerosis, Thrombosis, and Vascular Biology. 2004;**24**:31319

[115] Krishaswamy R, Devaraj SN, Padma VV. Lutein protects HT29 cells against Deoxynivalenol-induced oxidative stress and apoptosis: Prevention of NF-(B nuclear localization and downregulation of NF-(B and COX-2)) expression. Free Radical Biology & Medicine. 2010;**49**:5060

[116] Dilsiz N, Sahaboglu A, Yildiz M, Reichenbach A. Protective effects of various antioxidants during ischemia-reperfusion in the rat retina. Graefe's Archive for Clinical and Experimental Ophthalmology. 2006;**244**:62733

[117] Peng CL, Lin ZF, Su YZ, Lin GZ, Dou HY, Zhao CX. The anti-oxidative function of lutein: Electron spin resonance studies and chemical detection. Funct Plant Biology. 2006;**33**:83946

[118] Ashino T, Yamanaka R, Yamamoto M, Shimokawa H, Sekikawa K, Iwakura Y, et al. Negative feedback regulation of LPS-induced inducible NOS gene expression by HMOX-1 induction in macrophages. Molecular Immunology. 2008;**45**:210615

[119] Jin XH, Ohgami K, Shiratori K, Suzuki Y, Hirano T, Koyama Y, et al. Inhibitory effects of lutein on endo toxin-induced uveitis in Lewis rats. Investigative Ophthalmology & Visual Science. 2006;**47**:25628

[120] Kim JE, Leite JO, DeOgburn R, Smyth JA, Clark RM, Fernandez ML. A lutein-enriched diet prevents cholesterol accumulation and decreases ox-LDL and inflammatory cytokines in the aorta of

Guinea pigs. The Journal of Nutrition. 2011;**141**:145863

[121] Zou Z, Xu X, Huang Y, Xiao X, Ma L, Sun T, et al. High serum level of lutein may be protective against early ATS: The Beijing atherosclerosis study. Atherosclerosis. 2011;**219**:78993

[122] Hozawa A, Jacobs JDR, Steffes MW, Gross MD, Steffen LM, Lee DH. Circulating carotenoid concentrations and incident hypertension: The coronary artery risk development in young adults (CARDIA) study. Journal of Hypertension. 2009;**27**:23742

[123] Voutilainen S, Nurmi T, Mursu J, Rissanen TH. Carotenoids and cardiovascular health. The American Journal of Clinical Nutrition. 2006;**83**:126571

[124] Adluri RS, Thirunavukkarasu M, Zhan L, Maulik N, Svennevig K, Bagchi M, et al. Cardio-protective efficacy of a novel antioxidant mix vitaepro against ex vivo myocardial ischemia-reperfusion injury. Cell Biochemistry and Biophysics. 2011;**30**:93007

[125] Phaniendra A, Jestadi DB, Periyasamy L. Free radicals: Properties, sources, targets, and their implication in various diseases. Indian. Journal of Clinical Biochemistry. 2015;**30**:11-26 [CrossRef] [PubMed]

[126] Katerji M, Filippova M, Duerksen-Hughes P. Approaches, and methods to measure oxidative stress in clinical samples: Research applications in the cancer field. Oxidative Medicine and Cellular Longevity. 2019;**2019**:29. DOI: 10.1155/2019/1279250

[127] Rahman T, Hamzan NS, Mokhsin A, Rahmat R, Ibrahim ZO, Razali R, et al. Enhanced status of inflammation and endothelial activation in subjects with familial hypercholesterolaemia and their related unaffected family members: A case control study. Lipids in Health and Disease. 2017;**16**:81. DOI: 10.1186/s12944-017-0470-1

[128] Kohlmeier L, Kark JD, Gomez-Gracia E, Martin BC, Steck SE, Kardinaal AF, et al. Lycopene and myocardial infarction risk in the EURAMIC study. American Journal of Epidemiology. 1997;**146**:618-626

[129] Buijsse B, Feskens EJ, Schlettwein-Gsell D, Ferry M, Kok FJ, Kromhout D, et al. Plasma carotene and alpha-tocopherol in relation to 10-y all-cause and cause-specific mortality in European elderly: The survey in Europe on nutrition and the elderly, a concerted action (SENECA). The American Journal of Clinical Nutrition. 2005;**82**:879-886

[130] Ciccone MM, Cortese F, Gesualdo M, Carbonara S, Zito A, Ricci G, et al. Dietary intake of carotenoids and their antioxidant and anti-inflammatory effects in cardiovascular care. Mediators of Inflammation. 2013;**2013**:782137

[131] Gajendragadkar PR, Hubsch A, Maki-Petaja KM, Serg M, Wilkinson IB, Cheriyan J. Effects of oral lycopene supplementation on vascular function in patients with cardiovascular disease and healthy volunteers: A randomized controlled trial. PLoS One. 2014;**9**:99070

[132] Fortmann SP, Ford E, Criqui MH, Folsom AR, Harris TB, Hong Y, et al. CDC/AHA workshop on markers of inflammation and cardiovascular disease: Application to clinical and public health practice: Report from the population science discussion group. Circulation. 2004;**110**:554-559

[133] Blankenberg S, Rupprecht HJ, Bickel C, Peetz D, Hafner G, Tiret L,

et al. Circulating cell adhesion molecules and death in patients with coronary artery disease. Circulation. 2001;**104**: 1336-1342

[134] Blake GJ, Ridker PM. Inflammatory biomarkers and cardiovascular risk prediction. Journal of Internal Medicine. 2002;**252**:283-294

[135] van Herpen-Broekmans WM, Klöpping-Ketelaars IA, Bots ML, Kluft C, Princen H, Hendriks HF, et al. Serum carotenoids and vitamins in relation to markers of endothelial function and inflammation. European Journal of Epidemiology. 2004;**19**:915-921

[136] Kaulmann A, Bohn T. Carotenoids, inflammation, and oxidative stress--implications of cellular signaling pathways and relation to chronic disease prevention. Nutrition Research. 2014;**34**:907-929

[137] Di Tomo P, Canali R, Ciavardelli D, Di Silvestre S, De Marco A, Giardinelli A, et al. Beta-carotene, and lycopene affect endothelial response to TNF-alpha reducing nitro-oxidative stress and interaction with monocytes. Molecular Nutrition & Food Research. 2012;**56**:217-227

[138] Khansari N, Shakiba Y, Mahmoudi M. Chronic infl ammation and oxidative stress as a major cause of age-related diseases and cancer. Recent Patents on Inflammation & Allergy Drug Discovery. 2009;**3**:73-80

[139] Dietary guidelines for Americans. Home and Garden Bulletin no 232. 5th ed. Washington, DC: US Department of Agriculture, US Department of Health and Human Services; 2000

[140] Bendall JK, Cave AC, Heymes C, Gall N, Shah AM. Pivotal role of a gp91phox-containing NADPH oxidase in angiotensin II-induced cardiac hypertrophy in mice. Circulation. 2002;**105**(3):293-296

[141] Umar S, Van Der Laarse A. Nitric oxide and nitric oxide synthase isoforms in the normal, hypertrophic, and failing heart. Molecular and Cellular Biochemistry. 2010;**333**(1-2):191-201

[142] Lijun DAI, Brookes PS, Darley-Usmar VM, Anderson PG. Bioenergetics in cardiac hypertrophy: Mitochondrial respiration as a pathological target of NO*. American Journal of Physiology. 2001;**281**(6):H2261-H2269

[143] Saavedra WF, Paolocci N, John MES, et al. Imbalance between xanthine oxidase and nitric oxide synthase signaling pathways underlies mechanoenergetic uncoupling in the failing heart. Circulation Research. 2002;**90**(3):297-304

Section 5

Carotenoids and Biological Roles

Chapter 7

Carotenoids in Thermal Adaptation of Plants and Animals

Ivan M. Petyaev

Abstract

The support of carotenoids in photosynthesis is well documented. However, what is their role in parts of plants where there is no photosynthesis such as in fruits or stems or even in parts which are not exposed to the light at all, such as seeds or roots? Why are carotenoids essential for all animals and humans and present in almost every tissue in their body? The answer is that carotenoids can make complexes with lipids, which results in an increase of lipid thermal energy absorption and a reduction of viscosity. These changes help to expand the temperature range for the functionality of lipid structures, improve the capacity of thermal homeostasis and support adaptation and survival of living species to environmental stress and in particular to temperature variations. Working as "thermal antennas" carotenoids can increase lipid thermal energy conductivity, heat storage and heat retaining capacity. This, on the one hand, can reduce the freezing/melting points of plant and animal lipids and makes carotenoids work as antifreezers in microorganisms, plants or ectothermic animals. On the other hand, the thermal antennas can help absorb, transmit and accumulate external thermal energy essential to activate and support cellular metabolism. In addition, we describe how these properties of carotenoids can affect lipid parameters in nutrition, physiology and pathology.

Keywords: carotenoids, complex with lipids, thermal adaptation, heat storage and conductivity, lipid viscosity, cholesterol crystals, atherosclerosis, calorie intake

1. Introduction

Carotenoids are tetraterpenoid molecules, which play an important role in the photosynthesis of bacteria, algae, fungi and plants. However, these molecules may have other not always clear roles because they could be synthesised in fruit chromoplasts or in plant roots, which are not involved in light-dependent energy conversion. In animals and humans, who can get carotenoids only with food, these molecules have an additional functionality, for example, in the retina they facilitate filtration of light with different wavelengths.

Apart from light-harvesting and antioxidant activities, carotenoids can modulate lipid phase transition in cellular membranes, which affects their fluidity, polarity and, consequently, their functional properties [1, 2]. In contrast to membrane biology, the role of carotenoids in non-membrane lipid structures, such as lipid droplets or circulating lipoproteins, remains unclear. Only one property of these molecules is well documented, that they are hydrophobic antioxidants and can effectively protect lipoproteins and other lipids from free radical oxidation, which typically occurs in different pathological conditions [3, 4].

In this paper, based on our observations, we suggest that carotenoids may have another biological role. By creating thermodynamically favourable complexes with lipids, they can work as "thermal antennas" facilitating thermal energy absorption and conductivity, increase heat storage and heat retaining capacity and distribution within the lipid matrix of cellular and tissue structures. The reduction of lipid viscosity by carotenoids not only improves the control of thermogenesis but also increases oxygen solubility in lipid extra- and intra-cellular structures and boosts plasma lipoprotein oxygen transport, activation of mitochondria and tissue respiration.

2. Materials and methods

In this paper, we present the results of our experiments on the interaction of different carotenoids and lipids *in vitro*, *ex vivo* and *in vivo.* In these experiments, we tested the representatives of two main groups of carotenoids: highly hydrophobic carotenes—lycopene and β-carotene, and less hydrophobic ones, containing oxygen xanthophylls—lutein, zeaxanthin and astaxanthin. Since in most of the experiments interaction of different carotenoids and different lipids led, in principle, to similar patterns or trends, and due to the limitation of the space in this paper, we present here as an example only representative data from each set of undertaken experiments.

3. Product preparation

3.1 Saturated predominantly long-chain fatty acids—dairy butter

546 mg of lycopene (Lyc-O-Mato, LycoRed) was blended into molten 2340 g of unsalted dairy butter (President, France) at 45°C. The blend was stirred thoroughly for 10 min to ensure an even mixture, which was afterwards dispensed into 30 g individual moulds/containers by pouring. Each 30-g butter sample contained 7-mg lycopene. Storage in sealed containers was at −20°C up to 3 months and at +4–8°C for no more than 1 month.

3.2 Monounsaturated and polyunsaturated fatty acids—vegetable oils

70 mg of lycopene was blended into a 50 mL volume of olive oil (Extra Virgin, Napolitana) or sunflower oil (Flora). Once dispersed, carotenoid oil samples were stored at the ambient temperature of 20–21°C away from light.

3.3 Lycopene supplement formulation for clinical trial

Preparation of highly bioavailable GA lycopene was described in our earlier publication [5].

DOI: http://dx.doi.org/10.5772/intechopen.104537

4. Clinical trial

4.1 Study design

The studies were conducted at the Institute of Cardiology, the Ministry of Health of the Russian Federation (Saratov, RF). The protocol was approved by the local ethics committee. All patients were informed about the purpose of the study and had given written consent regarding their participation in the study.

Two-group studies were executed in a parallel, double-blind, controlled fashion for 4–8 weeks. Subjects visited the clinic for anthropometric and blood pressure measurements and for drawing blood samples. During the first visit to the clinic, subjects received an aliquot of study product sufficient for the first two study weeks. Then, subjects were provided with additional 14-day aliquots of study product during visits to the clinic after every two follow-up weeks.

Of the three-group or multi-group studies, two intervention groups were run at the same time, followed by the other group(s), also two at the same time.

4.2 Selection criteria

Study subjects were selected from the pool of patients based on the criteria for oxidative stress and having low-to-moderate hyperlipidemia.

The criterion for hyperlipidemia would be applied in most studies to facilitate the study of subjects with increased cardiovascular risk. Subjects were, nevertheless, not under medical treatment for the management of cardiovascular risk factors.

4.2.1 Inclusion criteria

- Caucasian male or female subjects 47–69 years old
- C-reactive protein (CRP) > 6 μg/mL (*if applicable)*
- Malondialdehyde > 40 μM/mL (*if applicable)*
- Untreated low-to-moderate hyperlipidemia (total cholesterol (TC) ≥ 250 mg/dL and low-density cholesterol (LDL) ≥ 150 mg/dL.)
- Willing to sign to informed consent
- Non-smokers or moderate smokers (≤ 10 cigarettes daily)
- No anti-hypertensive, hypoglycemic, lipid-lowering or any other cardiovascular drugs
- Willingness and ability to comply with the protocol for the duration of the trial.

4.2.2 Exclusion criteria

- Unwillingness to sign to informed consent
- Unable to comply with the protocol for the duration of the trial

- Significant medical condition that would impact safety considerations
- Significantly elevated liver-specific enzymes, hepatitis, severe dermatitis, uncontrolled diabetes, cancer, severe gastrointestinal disease, fibromyalgia, renal failure, recent cerebrovascular accident (CVA), pancreatitis, respiratory diseases and epilepsy
- Compulsive alcohol abuse (> 10 drinks weekly), or regular exposure to other substances of abuse
- Participation in other nutritional or pharmaceutical studies
- Positive test for tuberculosis, human immunodeficiency virus (HIV) or Hepatitis B.

4.3 Experimental procedure

Recruited study subjects were non-randomised into two intervention groups: a lycopene group (n=10) and a placebo group (n=10). This non-randomised method was chosen to ascertain the similarity of subject characteristics between study groups with respect to age, gender distribution and total cholesterol. Because the experimental groups were small (n=8–10), randomisation would likely increase the risk of variable group composition.

The Principal Investigator was provided with coded study product, according to the number of subjects in intervention(s) and control, and allocated an equal number of study subjects to either study product such that each group (corresponding to the same product code) matches the other group in numbers and demographic and clinical variables (as explained above). The Principal Investigator was blinded to which codes represented active or control product.

Prior to the start of the study, subjects were instructed to refrain from the consumption of dairy butter and tomato-based products for 10 days before beginning the trial and for the duration of the study. At the start of the study, following the end of the 10-day run-in period, blood samples were drawn and subjects were checked for meeting inclusion criteria. Then, subjects were distributed with a 2-week supply of the study product. Anthropometric measurements and blood pressure were taken and blood samples drawn.

All volunteers were instructed to ingest one sample of dairy butter once a day with the main meal.

All blood parameters were measured in the morning between 8 and 10 am.

Blood was collected in the morning after an overnight fast from arm veins of the volunteers. Serum was separated from the clotted mass by centrifugation and aliquots were stored at –80°C prior to analysis.

4.4 Compliance

All volunteers were instructed to ingest dairy butter sample once a day with the main meal. All volunteers were instructed to keep their butter packaging and bring it to the following clinical visit. After the verification of compliance, the packaging was exchanged for a fresh 14-day supply of products.

5. Methods

5.1 Microscopy—product samples

Microscopy was used to measure the size of lipid droplets and fat globules. For the visualisation of lipid droplets and fat globules and the quantification of their size, a binocular microscope Olympus BX41 was used with Cell^B software for morphometric analysis. All the parameters were collected from 10 randomly selected microscopic fields of 800 μ^2 at ×1000).

5.2 Microscopy—clinical samples

Typical skin smear samples were collected before the supplementation and 4 weeks after [5]. The duplicates of the specimens were fixed, stained and analysed: the first one for lycopene with fluorescein isothiocyanate conjugated monoclonal mouse antibodies, at 200× magnification, and the second for lipid droplets of the sebum with Oil Red O, at 1000× magnification, as described previously [6].

5.3 Melting and defrosting time

Approximately 20 mg of butter, or pork fat, or beef fat, or cocoa butter with or without certain concentrations of carotenoids, were placed on the surface of the laboratory slides and incubated at 37°C in a laboratory incubator (TLK39) until they melted.

To determine defrosting time, frozen oil aliquots of 200 mL were incubated at an ambient temperature of 20±2°C until they were completely melted. The period of time required for the tested sample to be melted or defrosted was measured with a laboratory timer (QUANTUM).

5.4 Cooking chicken breast

Chicken breast samples of 150 g were used. Before cooking, each sample was doused either in the pure olive oil (control) or in oil with a carotenoid. The oven was pre-heated to 275°C. A probe thermometer with a digital reading was placed in the thickest point of the sample. The sample was then placed in the oven and a reading taken and recorded every 300 sec for the 900-sec duration of the cooking.

5.5 Cooking salmon

Wild pacific Keta salmon fillets, 115 g each, were oven-cooked in individual small aluminium foil containers. The following preparation conditions were applied before cooking: either addition of 25 mL of the olive oil containing Lycopene 0.23 mg/mL) or control—addition of 25 mL of pure olive oil. Salmon fillets were turned several times making sure that they are completely covered with the added oil. After that, 500 mg of salt and 5 mL of fresh lemon juice were applied on the surface of each portion of fish. During the cooking process, the measurement of the internal (doneness) temperature of the fish was performed using a digital thermometer at the following time points: 8 min, 12 min, 16 min, and 20 min. All the containers had to be taken out of the oven for temperature measurements; hence, only the time when the fish was in the oven was counted as the cooking time. Once the internal temperature of the fish reached 62°C (doneness temperature for salmon), a small (about 1 g) fragment of fish was immediately taken from the relevant piece (using a scalpel blade and thumb forceps) and placed in a 15-mL laboratory tube containing 1 mL of distilled water. An additional sample of fish was also taken before cooking.

5.6 Salmon sample preparation

Samples were transferred to the laboratory and weighed using analytical scales (Discovery DV114C, OHAUS Corp.). Distilled water was added to each sample apart from the 'juice' and 'sauce' samples to provide the ratio of 9 mL of water per 1 g of sample (these samples were regarded as 1/10 dilutions). 1 mL of distilled water was added to the 'juice' and 'sauce' samples to produce 1/2 dilutions. Following this step, all the samples were homogenised using IKA T10 basic Ultra-Turrax homogeniser system at maximum speed (30,000 RPM). After the homogenisation of each sample, the homogeniser was disassembled, and both its rotor and stator were carefully cleaned in order to prevent sample cross-contamination. Following homogenisation, second dilutions with four volumes of distilled water (1-mL homogenate + 4 mL of water) were prepared from the homogenates of the raw fish, fish cooked with water and fish cooked with either pure olive oil, or containing lycopene, or astaxanthin (resulting in 50× dilutions). Similar dilutions were made from salmon 'juice' and 'sauce' samples described above (resulting in 10× dilutions). Finally, additional fivefold dilutions were made from the 50× dilutions of homogenates prepared from all salmon samples (resulting in 250× dilutions).

5.7 Vitamin B12 concentration determination

All samples were analysed within 48 hours following cooking experiments. Vitamin B12 BioAssay™ ELISA Kit (US Biological) was used for Vitamin B12 concentration determination. Vitamin B12 concentrations in the samples from the cooking experiment were determined in both undiluted supernatants (1 g of liver + 19 mL of buffer) and dilutions 1/2, 1/4 and 1/8 (the latter only for the samples cooked without lemon juice). The dilutions were prepared using sample dilution buffer (phosphate-buffered saline—PBS) supplied with the kit. B12 concentration evaluation was performed in 50 μL of solution according to the protocol provided with the kit.

Vitamin B12 concentration determination was performed using Multiscan FC microplate photometer (Thermo Fisher Scientific) by measuring optical light absorbance at 450 nm (reference wavelength 620 nm) as recommended by the kit manufacturer. All the calibration standards were measured in duplicates. Measurement results were analysed using SkanIt software for Multiscan FC system (a four-parameter logistic algorithm was applied). Vitamin B12 concentrations in the original samples were obtained by re-calculation taking into account sample dilutions during material processing. Once all the measurements were completed, the results of the two cooking experiments were combined by taking average concentration value for each set of conditions.

5.8 Formation of cholesterol crystals

A solution of 198 mL of 1 g of cholesterol (Sigma) in 99% ethanol was divided in two equal parts. In one part, 1 mL of ethanol containing 100 μg of the dissolved lycopene was added. This provided a ratio of lycopene to cholesterol as $1:10^6$.

Into the other part, the control, 1 mL of the ethanol itself was added.

Then, both the samples were left in a dark room for evaporation under 20–22°C.

Recording of the status of both the samples was made at least daily or at even shorter intervals.

5.9 Disassembly of cholesterol crystals

20 mL of ethanol solutions with different concentrations of lycopene were added to 100 mg of crystallised cholesterol. As a control, 20 mL of the same ethanol was used but without any lycopene. After gentle stirring for a couple of minutes, at the room temperature of about 20–22°C, recording of the results was made.

5.10 Ex vivo experiment with cholesterol crystals in the arterial wall

It was important to try to check whether carotenoids, and lycopene in particular, could affect the folding of cholesterol crystals, which are not just synthesised by a manufacturer, but produced naturally, and particularly those, which are developed during pathological process in human. For this purpose, we used pieces of atherosclerotic abdominal aorta, which were obtained, during a combined vascular graft and bypass surgery.

Comparable types of atherosclerotic lesions were collected. This was in terms of their stages of development, with prominently featured cholesterol crystals of similar size, embedded into the atheromatous tissues of the aorta wall. As a control material, we collected the pieces of atherosclerotic abdominal aorta containing calcium phosphate crystals.

First, an ethanol solution 1 μg/mL of lycopene was prepared. It was then diluted by PBS 10-fold. As a control, solution was made with the same ratio ethanol to PBS but without lycopene. Then, the pieces of aorta were incubated, in light-protected containers at a room temperature of about 20–22°C for 13 days.

5.11 Clinical parameters

The body mass index (BMI) of the participants was measured in the morning and calculated in kg/m^2. Pulse rate and systolic and diastolic blood pressure were measured three times in the left arm of the seated patient after 15 min of rest. The time between measurements was no less than 2 min. The mean number for each parameter was calculated.

All the body and vascular parameters were measured in the morning between 8 and 10 am.

5.12 Blood biochemistry

Biochemistry and inflammatory markers, glucose, total cholesterol (TC), triglycerides (TG), high-density cholesterol (HDL), low-density cholesterol (LDL), C-reactive protein (CRP) and oxidised LDL (LDL-Px) were measured using commercially available analytical kits according to manufacturers' instructions (BioSystems, Medac, R&D Systems). Inflammatory oxidative damage (IOD—malondialdehyde) was measured applying a colorimetric method [7].

5.13 Statistics

For the assessment of normally distributed parameters, the Shapiro-Wilkinson method was used. Student's t-test was then applied for both the paired and unpaired samples.

In cases where parameters were not normally distributed, the Mann-Whitney U test and the Kruskal-Wallis test were used.

The analysis of variance (ANOVA) and the analysis of covariance (ANCOVA) were used with post hoc analysis (Statistica9 suit, StatSoft; Inc.). Statistical significance between two-tailed parameters was considered to be $p<0.05$.

6. Results

6.1 Carotenoid-lipid interactions

The integration of carotenoids and lipids resulted in a number of changes in the light absorption spectrum of the former, subject to the specific nature of the interacting molecules. For example, there was a red shift in the visible part of absorption of lycopene when it was added into the sunflower oil (**Figure 1a**) or a hyperchromicity in the spectrum of lutein when it was added to the cocoa butter (**Figure 1b**). These changes indicate that the carotenoids were able to interact with the lipid molecules and created physical complexes with them, which would be thermodynamically more favourable than when all these molecules were present in free forms.

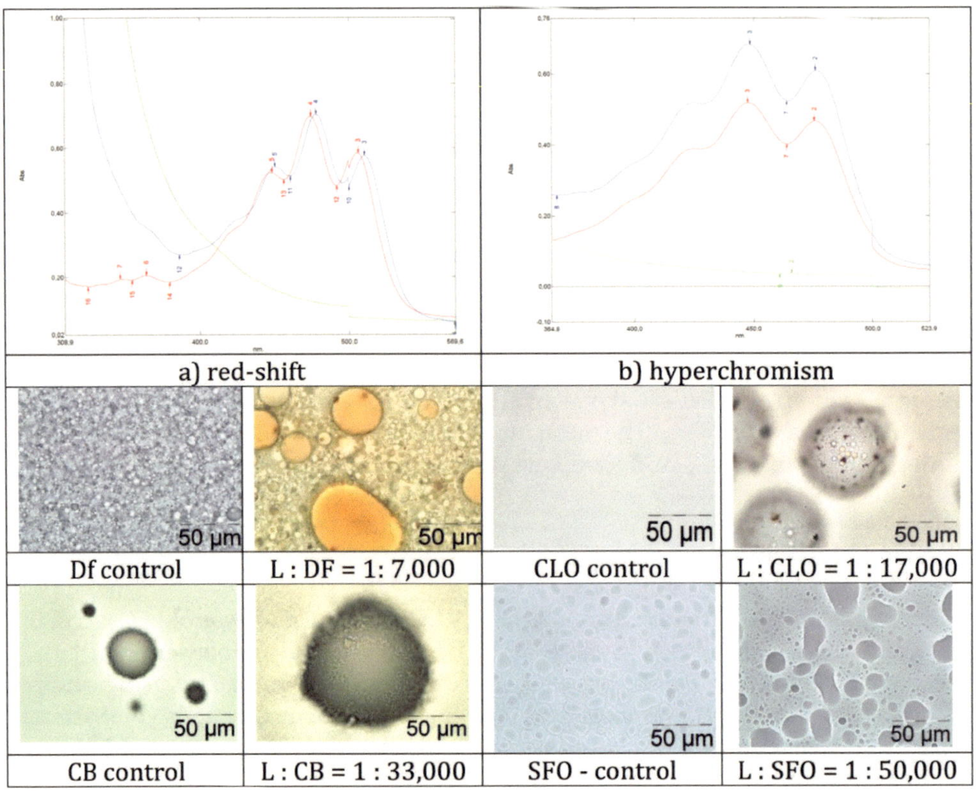

Figure 1.
Formation of thermodynamically favourable complexes between carotenoids and lipids. a) Red shift in light absorption peaks of lycopene embedded into sunflower oil, in 1:330 m/m—blue, control lycopene—red, control sunflower oil—green. b) Hyperchromism in light absorption peaks of lutein embedded into cocoa butter, in 1:330 m/m—blue, control lutein—red, control cocoa butter—green; in all experiments above, ethanol-methylchloride 5:1 w/w was used. Typical microscopy slides at 1000× magnification: DF—dairy fat, CLO—cod liver oil, CB—cocoa butter, SFO—sunflower oil, L—lycopene, m/m.

To visualise how formation of these complexes affects lipids and subsequent changes in their structures, we used light microscopy. It was found that the incorporation of carotenoids into oils or fats matrix may result in a dose-dependent growth of the size of lipid droplets or fat globules. This was observed in a broad range of plant or animal oils and fats. The working concentration at which a particular carotenoid could trigger this microscopy-visual effect was dependent on the structure of carotenoid molecules and on the length and saturation level of fatty acids (FA) in the lipids used. Overall plant lipids with a higher level of unsaturated FA were more sensitive to carotenoid-induced changes than animal lipids. Within the first group, vegetable oils with longer FA were more responsive than nut oils or cocoa butter. Within animal lipids, fish oils with a higher level of unsaturated and longer FA were more responsive than the bird fats, and they, in turn, were more sensitive than mammalian fats (**Figure 1**, microscopy slides).

For example, one molecule of lycopene was able to make noticeable changes in the size of the lipid droplets of the dairy fat/butter starting with a ratio 1:30,000 of molecules of its triglycerides. For the cod liver oil, this threshold was lower at one molecule of lycopene per about 40,000 molecules of the oil, for the cocoa butter, it was at 1 per 80,000 and for the sunflower oil, it was at 1 per about 120,000 molecules of their triglycerides.

6.2 Carotenoids reduce lipid viscosity *in vitro* and *in vivo*

The formation of complexes between carotenoids and lipids resulted in the reduction of their viscosity. Plant oils response was more significant than the animal fats. For example, one molecule of lycopene added to about 67,000 molecules of olive oil

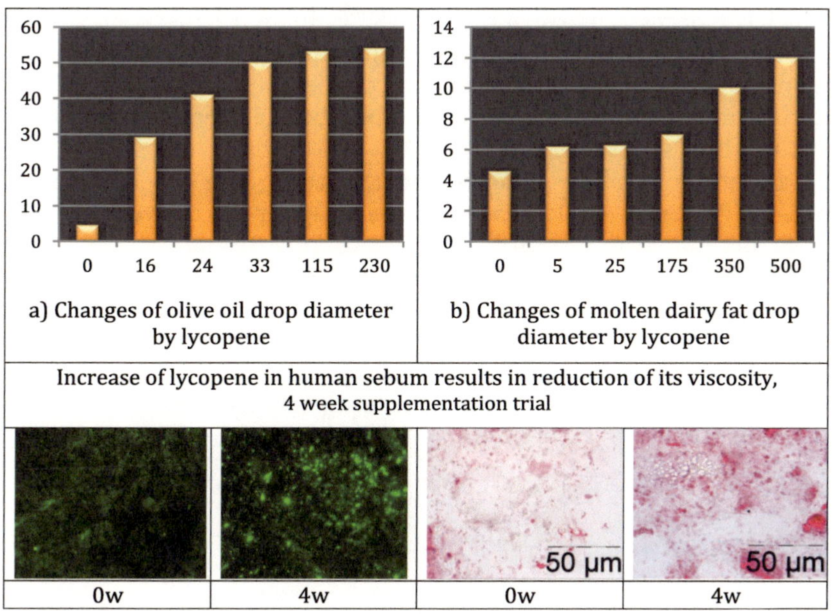

Figure 2.
Carotenoids reduce viscosity of lipids in vitro top slides and in vivo bottom slides. Viscosity was measured as a diameter, y-axis in μm, of a single drop of the oil on the surface of the water (a) or a single drop of the molten butter on the hard surface (b) at 37°C, before the experiment the butter and oil samples were warmed to that temperature too; x-axis—lycopene concentration in μ/g; each column is an average of three independent measurements. The microscopy of typical skin smear samples from 60-year-old clinically healthy man, collected before and 4 weeks after supplementation with 7 mg of highly bioavailable GA lycopene [5].

lipids could reduce viscosity of the oil, in terms of its drops spreadability, by 10-fold, or 1000%; when it was added to about a similar amount of dairy fat lipid molecules, the reduction of the viscosity was only by 50% (**Figure 2a** and **b**).

Carotenoids could change the viscosity of lipids not only when they were incorporated in their matrix *in vitro*, but *in vivo* as well. For example, when middle-aged persons, who were clinically healthy but had age-associated lycopene functional deficiency, were supplemented with this carotenoid for 4 weeks, the viscosity of their sebum, in terms of the diameters of the lipid droplets collected from the surface of the skin, was significantly increased (**Figure 2**, two slides on the bottom-right). The fact that these changes were not just associated with the lycopene intake but caused by its accumulating in these droplets was confirmed by the direct measurement of this carotenoid in the collected sebum (**Figure 2**, two slides on bottom-left).

6.3 Carotenoids increase thermal energy absorption, heat storage capacity, heat retention and thermoconductivity of lipids

The incorporation of carotenoids into plant or animal lipids increases their rate of thermal energy absorption, the amount and the time of this energy storage. For example in **Figure 3**, when the same level of heat was applied, the sunflower oil with carotenoids could start to accumulate this heat faster and become hotter by 5°C, and after the external heat was switched off, the retention of the heat lasted significantly longer.

The increase in lipid thermoconductivity by carotenoids was demonstrated in another set of experiments. For oils, liquid at room temperature, we froze them first and then measured this parameter in terms of time, which was necessary to defrost these oils. For lipids, solid at room temperature, we assessed thermoconductivity as a time, which was necessary to melt them at +37°C. In these experiments, carotenoid increase in thermoconductivity was more prominent in plant oils than in animal fats.

For example, the same concentration of lycopene, 330 μg/mL, could reduce the defrosting time of the olive oil by 12-fold, but the cod liver oil only by twofold (**Figure 4a** and **b**).

For lipids, solid at room temperature, the same as in above experiments, the concentration of lycopene could increase the melting time of cocoa butter by more than 10-fold, but for dairy fat by only threefold (**Figure 4c** and **d**).

This increase in thermoconductivity could also be observed in the heating not just in lipids but in their emulsions in water too. For example, lycopene in concentration

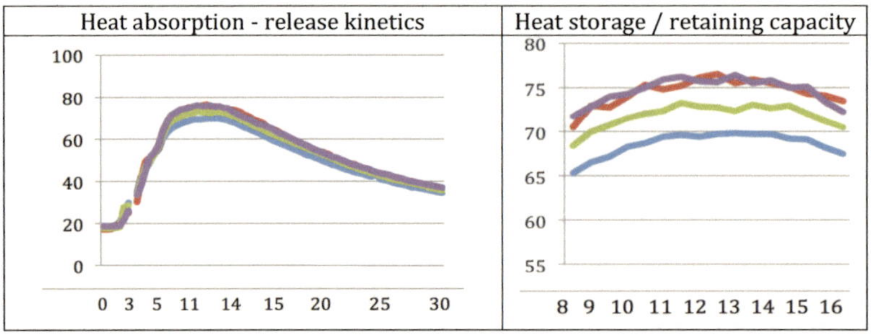

Figure 3.
Carotenoids increase the heat storage capacity of sunflower oil. Vertical axis—temperature t °C/mL, horizontal axis—time of the experiment, in min; blue—control oil, green—with 9.3-μM lycopene, purple—with 9.3-μM β-carotene, red—with18.6-μM astaxanthin.

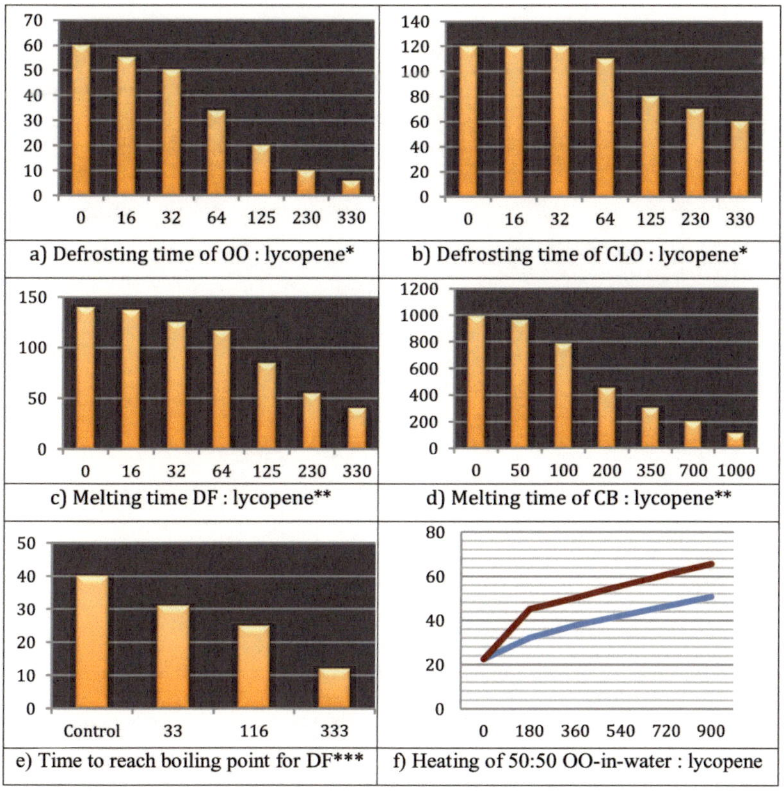

Figure 4.
Carotenoids increase thermal energy conductivity in different lipids and their emulsions. OO—olive oil, other abbreviations as in **Figure 1**. **Defrosting time from −30°C of frozen oils to room temperature, x-axis—lycopene concentration in μg/mL; in kerosene, astaxanthin and lycopene concentrations were 16μg/mL. **Melting time at +37°C to room temperature, x-axis in this and the following experiment—lycopene concentration in μg/g. ***Boling was at +350°C; in all the above experiments, y-axis is time in seconds, and each column is an average of three independent measurements. Kinetics of heating of OO-in-water emulsion, at heated 150°C stove, y-axis—temperature in °C, x-axis—time in seconds, blue line—control, red—with lycopene in 230 μg/mL.*

of 330 μg/g could accelerate the time to reach the boiling point for dairy butter by fourfold (**Figure 4e**). In another experiment, to reach a temperature from 20°C to +45°C for 50:50 olive-oil-in-water emulsion, with the same carotenoid in concentration of 230 μg/mL, took only 3 min, when for the control emulsion, it was 7 min. Moreover, carotenoids were not just able to increase the rate of heating but the maximum level of the temperature the emulsions could reach. In this particular experiment, the increase was by 14°C, from 51°C for the control emulsion to 65°C when lycopene was therein (**Figure 4f**).

6.4 Carotenoid increase in heat absorption of oils accelerates the cooking process and renders it healthier

Increase in thermoconductivity of oil doped by carotenoids may result in an accelerated cooking process. For example, to reach the 84°C temperature of completed cooking for a piece of chicken breast, coated in olive oil at an oven temperature of 180°C, took 15 min. When this oil contained lycopene or astaxanthin, in concentration 0.23 mg/mL, cooking took 13 and 10 min, respectively (**Figure 5**). For a fillet of wild salmon to

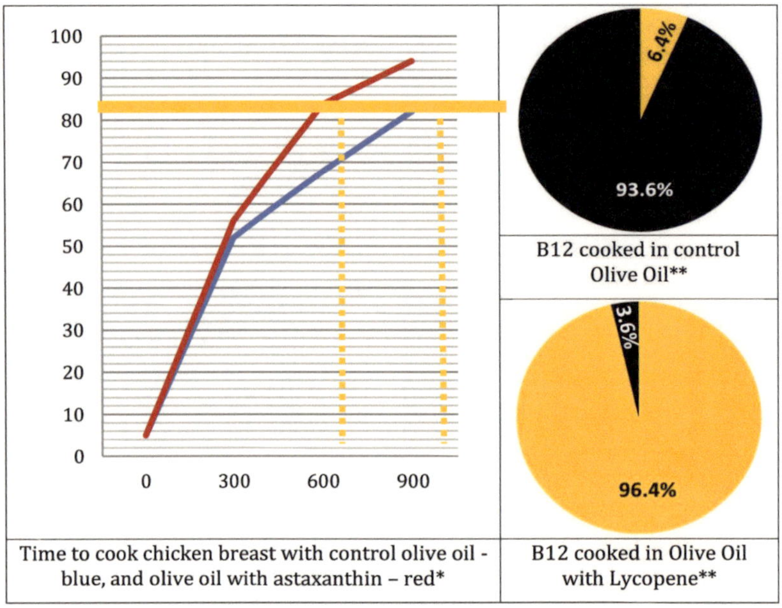

Figure 5.
Carotenoids in olive oil accelerate cooking time of chicken breast and preserve vitamin B12 in cooked salmon.
**Astaxanthin concentration was 0.23 mg/mL in olive oil; y-axis—temperature in °C, x-axis—time in seconds.*
***Yellow—content of preserved intact vitamins, black—lost vitamins.*

reach the temperature of completed cooking at 62°C, it took 9 min for the samples in the control oil and 7 or even 6 min for the oil with lycopene or astaxanthin, respectively.

The observed acceleration of the cooking process and, consequently, reduction in the cooking time may help in preservation in the cooked food of important thermo-sensitive macronutrients or vitamins. For example, in the above experiment, the concentration of vitamin D3 in the baked salmon was only about one-third of its pre-cooked level. However, when oil contained astaxanthin or lycopene, the remaining level of this vitamin was significantly higher, more than 55 or 82%, respectively. In the same fish, samples of Vitamin B12 were more sensitive to the cooking temperature; by the end of the experiment, only about 6% of these molecules were detected there. Using the oil with astaxanthin did not make much difference, but with lycopene saved more than 96% of this vitamin in the cooked fish (**Figure 5**).

6.5 Carotenoids reduce the digestion rate of lipids and their absorption level

The industrial process of refining oil production removes all its ingredients, including carotenoids, which are originally present in freshly pressed plant oils and fats. As a result of this, the lipid droplets of these oils have higher viscosity, are significantly smaller, and have a faster digestibility rate leading to an increase in calorie release and absorption. As a result of this, refined vegetable oils alongside refined sugars are the main food sources of easily released calories, the main dietary factor contributing to the development of the global obesity pandemic.

The reintroduction of carotenoids to plant oils and fats increases the size of their droplets or globules (**Figure 1**). Consequently, the time of digestions of these lipids will increase, and subject to the ingested lipid volume, not all lipids would be digested and absorbed. This would lead to a reduction in the amount of the absorbed lipid in the postprandial blood.

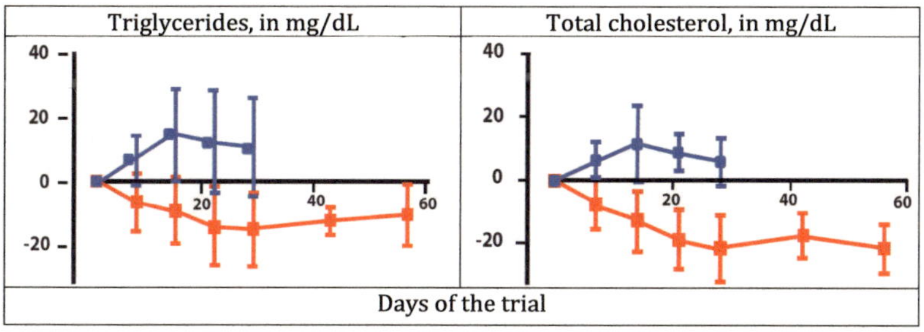

Figure 6.
Effect of daily ingesting of dairy butter with or without lycopene on the level of serum lipids in the fasting blood of volunteers. Blue—30 g of control dairy butter, red—30 g of the butter with 7-mg lycopene.

This was confirmed in a crossover clinical trial on healthy volunteers, who were asked to ingest different unmodified fat or fat-rich food products and, after one week's rest, ingest the same products but modified by carotenoids. Postprandial blood was analysed to assess the level and kinetics of the absorbed lipids. For example, after ingestion of 50 g of control dairy butter, area under the curve (AUC) for the first 4 hours for serum total cholesterol was 33±3.9 mg/dL and for triglycerides 25±2.7 mg/dL (n=10). However, the AUC after ingestion of the same amount of butter with 7 mg of lycopene and these parameters were reduced for cholesterol significantly to 22±2.5 mg/dL ($p<0.05$) and for triglycerides only as a trend to 19±2.2 mg/dL ($p>0.5$).

After repeating this experiment with 50 mL sunflower oil, with a slice of white bread, the effect of ingestion of lycopene-modified fat was even stronger. In the control experiment, the AUC for serum total cholesterol was 75±9.2 mg/dL and for triglycerides 56±6.8 mg/dL (n=10). The ingestion of the same amount of this oil but with 7 mg of lycopene resulted in a significant reduction of these parameters to 32±4.3 mg/dL ($p<0.01$) and 12±3.9 mg/dL ($p<0.001$).

In the next set of experiments on clinically healthy persons, with borderline hyperlipidaemia, we demonstrated that regular, daily intake of dairy butter with incorporated lycopene could reduce serum triglycerides, total cholesterol and LDL (**Figure 6**).

The butter trial was planned for 2 months. However, its control group was terminated earlier on ethical grounds because there was a significant rise in both blood lipids. In the lycopene butter group, at the end of the trial, the reduction of triglycerides was by 10 mg/dL and total cholesterol by about 20 mg/mL.

The daily ingestion of lycopene chocolate resulted in a significant reduction of both lipids, while in the groups which ingested either the same amount of control chocolate or lycopene in a capsule, there were no changes in these parameters.

6.6 Carotenoids reduce the rate of formation cholesterol crystals and facilitate their dissolutions

The ability of carotenoids to create thermodynamically favourable complexes with lipids, which changes their crystalline properties, can also be observed on their interactions with cholesterol. It was observed that the addition of carotenoids could significantly reduce the rate of cholesterol crystallisation. For example, in the experiment described in the legend to **Figure 7a**, visible cholesterol crystals in the control solution started to appear in 24 hours from the start of the evaporation of the solvent. When lycopene was introduced in a ratio of 1:1000 molecules of cholesterol, it took five times longer before these crystals started to be observed.

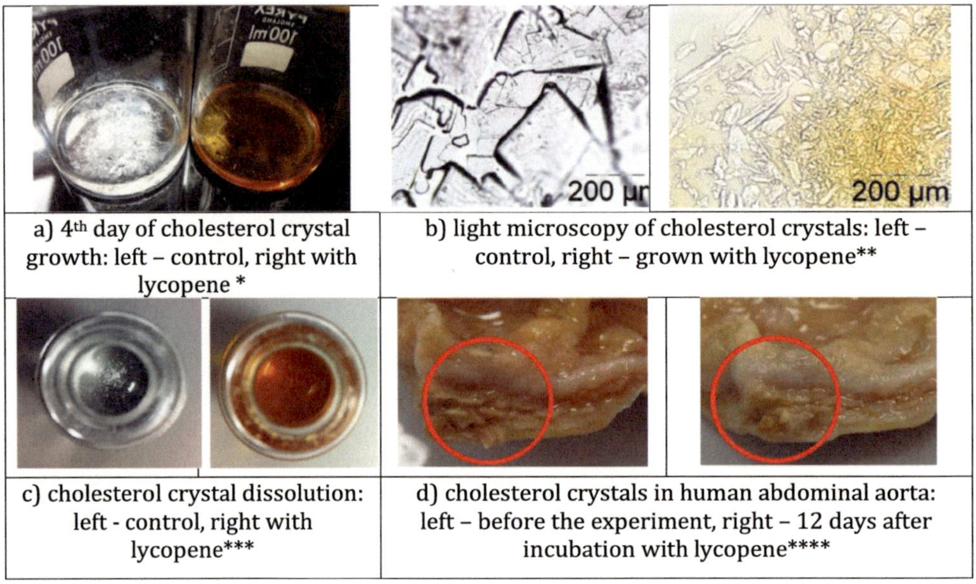

Figure 7.
*Lycopene and cholesterol crystals: reduction of their growth rate (a), disruption of their folding (b) and facilitation of their dissolution in vitro (c) and ex vivo (d). *2.5 mM of cholesterol in ethanol without or with 2.5-μM lycopene were evaporate at room temperature in a dark room. **Molecular ratio lycopene to cholesterol 1:4000; ***5 mM of cholesterol in ethanol without or with 2.5-μM lycopene. ****Incubation at room temperature in a dark room in the PBS solution with pre-diluted in ethanol lycopene in concentration 0.1 μg/mL, with NaN_3 to prevent microbial growth.*

It was interesting that it was not just the reduction in the rate of crystallisation we observed but also a new type of crystals emerged: they were significantly smaller, and some had needle forms in contrast to the much bigger slab-shaped crystals of unmodified cholesterol (**Figure 7b**). These observations confirmed that that lycopene, like other carotenoids, could create physical complexes with this type of lipids, disrupt their folding, clusterisation and affect their crystal structures.

This ability of carotenoids to disrupt folding in already existing cholesterol crystals was observed in our *in vitro* and *ex vivo* experiments. For example, lycopene could dissolve or help ethanol to dissolve cholesterol crystals in a mole ratio of 1:4000 or below (**Figure 6c**). Lutein was able to do the same, although in a lower concentration range (data not presented).

In the *ex vivo* experiment, a piece of abdominal aorta with massive atherosclerotic lesions and a combination of cholesterol and calcium phosphate crystals were incubated in phosphate-buffered saline containing lycopene pre-dissolved in ethanol. After 12 days of this incubation, not only a substantial amount of fat deposits of atherosclerotic plaques was dissolved, but also the number and size of cholesterol crystals were significantly reduced (**Figure 7d**).

At the same time, this incubation did not affect either the number or the size of calcium crystals in this piece of aorta or cholesterol crystals from a similar type of atherosclerotic lesion (data not presented).

7. Discussion

Our experiments demonstrated that the interaction of carotenoids with lipids could result in the formation of their complexes, which were thermodynamically more favourable than when these two groups of molecules were separate. As a result of this, carotenoids were better able to absorb the light energy in the longer wavelength of the red part of the spectrum.

The physical properties of lipids in these complexes were also changed due to a possible reduction in their cohesiveness between these molecules. For oil droplets or fat globules, this resulted in a reduction of their surface tension and, consequently, facilitated their fusion or enlargement. The larger the lipid droplets are, the less energy they have. Therefore, carotenoids can create complexes with lipids, which trigger their transition into a thermodynamically more favourable and stable phase. The fact that a single molecule of the carotenoid can affect the behaviour of 10,000 or 100,000 molecules of lipids may imply the possibility that the released energy, after formation of this type of complex, can dissipate beyond its physical location and cause a long-range transition of the lipid matrix, with consequent physical changes of its properties.

Lipid droplets with increased size would have less friction between each other, and the oils and fats became less viscous. Whether the reduction in the viscosity was a result of blending carotenoids in existing oil or fat products, or their incorporation into *de novo* assembled lipid structures, the effect in principle was the same. This was confirmed in the clinical trial, when the increase in lycopene concentration in the skin sebum resulted in the reduction of the viscosity of its droplets (**Figure 2**). These changes in sebum quality were similar to the observations on the supplementation of volunteers with other carotenoids such as lutein (results not presented here) and astaxanthin [6].

A significantly higher response in changes of the viscosity of plant lipids, over animal ones, could probably be either a result of their stronger interaction with carotenoids or easier spreadability of conformational changes in the plant lipids than in the animal fats. Or this can be a combination of both of these factors. Animals cannot synthesise carotenoids but plants can, and this is probably why the affinity to these molecules to other plant molecules, lipids, is higher than to animal ones.

The increased surface area of enlarged oil droplets or fat globules would have more surface-to-surface contacts with each other, which would facilitate the transfer of changing temperature between these particles. Whether there is a thawing of frozen liquid hydrocarbons, or melting of solid lipids, or their heating, the incorporation of carotenoids into these molecules accelerated temperature energy transfer within them.

Since one molecule of a carotenoid was able to facilitate temperature changes in 10,000, or 100,000 or more molecules of lipids, it is unclear whether this was due to changes in lipid molecule thermoconductivity, or carotenoids, in their complex with lipids, may work as 'thermal antennas', which could dissipate and distribute thermal radiation/energy far beyond the physical location of carotenoid-lipid complex.

The ability of carotenoids to increase the accumulation of thermal energy and its distribution was significantly higher for the plant lipids than for the fish oil, which, in turn, responded better than the mammalian fats. The reason for this effect of carotenoids on heat absorption gradient in different lipids could be the fact that plants are exposed to much broader variations in environmental temperature changes than ectotherms/poikilotherms like fish, when the body temperature of endotherms like mammals is constant.

The viscosity of lipids is the essential parameter, which controls cellular membrane permeability to electrolytes and nutrients, energy synthesis, cell growth and proliferation. One of the main factors determining the viscosity of lipids, and their ability to conduct the heat or the cool, is a lipid composition, a ratio of triglyceride saturated and unsaturated fatty acids, their length, other incorporated lipids, etc.

Our experiments indicate that carotenoids could have a new biological role not only to control viscosity of lipids but their thermal energy absorption, retention and conductivity too. If this is the case, this could be a much more efficient pathway to control these parameters. For a plant cell, to synthesise one molecule of a carotenoid, which can change the viscosity of 10,000 or even 100,000 molecules of lipids, would be much faster and more economic than to activate a lipid replacement process, which would involve a few hundred or thousand more new lipid molecules to be synthesised.

This possible new role of carotenoids as a factor facilitating adaptation to environmental, and in particular, temperature variation stresses, may explain a number of observations, which do not have clear explanations. What is the role of carotenoids, which are not involved in photosynthesis whether in a plant or in a light harvesting microorganism? What is the role of these molecules in parts of the plant where photosynthesis is not happening at all, like fruits or roots?

Since plants, or microorganisms such as algae, have exposure to much higher day-night, seasonal or other environmental temperature variability than animals, it is not surprising that the level of carotenoids in their tissue is 10^3–10^6 higher than in animals [8, 9]. Within animals, ectotherms, which do not have their own mechanism to control their temperature, rely more on the accumulation of ingested carotenoids than endotherms, which can maintain their thermal homeostasis. It is not surprising that in tissues of fish or reptiles, carotenoid concentration could be from 10- to 100-fold higher than in mammalians [10, 11].

The ability of carotenoids to work as antennas facilitating transmission and distribution of the thermal energy within lipid matrixes may find its practical applications in different industries. This carotenoid property may improve the performance of lipids, and possibly other hydrocarbons, when they are used for the production of greases, lubricants, liquid crystal devices, nanotubes, thermal energy storage, biodiesel and some other products, oils and fuels.

For example, carotenoids can accelerate the heating of oil or lipid-in-oil emulsions in general (**Figure 4f**), or when they used for cooking purposes in particular. This can shorten the time of the cooking process, save fuel, which is used to generate heating energy, and preserve more thermo-sensitive vitamins and micronutrients in the finished cooked meal (**Figure 5**). The correlation between the rate of acceleration of the heating of the cooking meal, in oils with different carotenoids, and the level of preserved thermo-sensitive vitamins was not always observed. This was probably due to additional antioxidant properties of carotenoids, which could contribute to preservation of these vitamins in the cooking process.

Another useful application of the ability of carotenoids to disrupt lipid folding would be to increase the size and reduce the viscosity of oil droplets and fat globules in food products. We demonstrated that the ingestion of dairy butter, vegetable oils and chocolate with enlarged lipid particles resulted not only in the reduction of the postprandial lipidaemia but also, if these products were regularly consumed, in the reduction of elevated fasting blood lipids. This means than carotenoids can be used to convert edible oils and fats into lipid lowering and weight management food products.

Perhaps, the culinary practice of cooking in oils/fats with lycopene-rich tomato sauce is a contributing factor as to why Italians are one of the slimiest nations in Europe and the USA [12, 13].

In our *in vitro* and *ex vivo* experiments, on the effect of lycopene and lutein of formation and dissolution of cholesterol crystals, we used a range of the ratio between these molecules the same as is present in humans from 1:4000 to 1:1000. The observed ability of carotenoids to interact with cholesterol molecules, to reduce the rate of its crystallisation, may be used for the prevention or control of the growth of cholesterol crystals and/or for the treatment by facilitating the disassembly of the already formed crystals, which are responsible for the rupture of atherosclerotic plaques leading to heart attack or ischaemic stroke [14, 15].

To assess the industrial, nutritional and medical applicability of using carotenoids to affect lipid properties, to create new materials, food and health care products would require more work and expertise in different fields.

In conclusion, it should be said that the main objective of the presentation of the data in this paper is to illustrate the new phenomenon, its potential biological role and practical applicability. The main body of the backing/supporting experiments on different carotenoids, lipids, doses, conditions, products, clinical trial participants, etc., would be a subject of future separate publications.

Acknowledgements

The author is grateful to his colleagues for conducting experiments results of which are presented in this paper: Dr Alexandre Loktionov Jr., Dr Alexey Petyaev, Professor, Pavel Dovgalevsky, Professor Viktor Klochkov, Dr Natalya Chalyk, Dr Dmitry Pristenskiy, Dr Marina Chernyshova, Dr Alexandre Loktionov, Dr Tatyana Bandaletova, Nigel Kyle, Marina Lozbiakova. A special thank you to Mr. William George for his personal supporting this project, Professor Yuriy Bashmakov for stimulating discussions in the filed, and Mrs. Anne George for editing this manuscript.

Conflict of interest

The author is a founder and director of Lycotec Ltd, the company that research and develop carotenoid based technologies.

Author details

Ivan M. Petyaev
Lycotec, Cambridge, UK

*Address all correspondence to: petyaev@aol.com

References

[1] Llansola-Portoles MJ, Pascal AA, Robert B. Electronic and vibrational properties of carotenoids: from *in vitro* to *in vivo*. Journal of the Royal Society Interface. 2017;**14**:20170504

[2] Wiesyaw IG, Strzayka K. Carotenoids as modulators of lipid membrane physical properties. Biochimica et Biophysica Acta. 2005;**1740**:108-115

[3] Young AJ, Gordon LL. Carotenoids—Antioxidant properties. Antioxidants. 2018;7:28. DOI: 10.3390/antiox7020028

[4] Griffiths K, Aggarwal BB, Singh RB, Buttar HS, Wilson D, De Meester F. Food antioxidants and their anti-inflammatory properties: A potential role in cardiovascular diseases and cancer prevention. Diseases. 2016;**4**:28. DOI: 10.3390/diseases4030028

[5] Ivan MP, Dovgalevsky PY, Klochkov VA, Chalyk NE, Pristensky DV, Chernyshova MP, et al. Effect of lycopene supplementation on cardiovascular parameters and markers of inflammation and oxidation in patients with coronary vascular disease. International Journal of Food Sciences and Nutrition. 2018;**6**(6): 1770-1777. DOI: 10.1002/fsn3.734

[6] Chalyk NE, Klochkov VA, Bandaletova TY, Kyle NH, Ivan MP. Continuous astaxanthin intake reduces oxidative stress and reverses age-related morphological changes of residual skin surface components in middle-aged volunteers. Nutrition Research. 2017;**48**:40-48

[7] Petyaev IM, Chalyk NE, Klochkov VA, Pristensky DV, Chernyshova MP, Kyle NH, et al. Pharmacokinetics and Oxidation Parameters in Volunteers Supplemented with Microencapsulated Docosahexaenoic Acid. International Journal of Applied and Basic Medical Research. 2018;**8**(3):148-154

[8] Bozalan NK, Karadeniz F. Carotenoid profile, total phenolic content, and antioxidant activity of carrots. International Journal of Food Properties. 2011;**14**:1060-1068

[9] Saini RK, Zamany AJ, Keum Y-S. Ripening improves the content of carotenoid, a-tocopherol, and polyunsaturated fatty acids in tomato (Solanum lycopersicum L.) fruits. 3 Biotech. 2017;7:43. DOI: 10.1007/s13205-017-0666-0

[10] Costantini D, Dell'Omo G, Casagrande S, Fabiani A, Carosi M, Bertacche V, et al. Inter-population variation of carotenoids in Gala´pagos land iguanas (Conolophus subcristatus). Comparative Biochemistry and Physiology, Part B. 2005;**142**:239-244

[11] Teglia CM, Attademo AM, Peltzer PM, Goicoechea HC, Lajmanovich RC. Plasma retinoids concentration in Leptodactylus chaquensis (Amphibia: Leptodactylidae) from rice agroecosystems, Santa Fe province, Argentina. Chemosphere. 2015;**135**:24-30. DOI: 10.1016/j

[12] Barrett AM, Boye KS, Oyelowo O. Burden of obesity: 10-year review of the literature on obesity prevalence in nine countries. In: ISPOR 13th Annual International Meeting. Toronto, Ontario, Canada; 2008

[13] Obesity Update 2017. Available from: www.oecd.org/health/obesity-update.htm

[14] Abela GS, Aziz K, Vedre A, Pathak DR, Talbott JD, DeJong J. Effect of cholesterol crystals on plaques and

intima in arteries of patients with acute coronary and cerebrovascular syndromes. The American Journal of Cardiology. 2009;**103**:959-968

[15] Vedre A, Pathak DR, Crimp M, Lume C, Koochesfahani M, Abela GS. Physical factors that trigger cholesterol crystallization leading to plaque rupture. Atherosclerosis. 2009;**203**:89-96